Design of Semiconductor QCA Systems

For a complete listing of titles in the
Artech House Nanoscale Science and Engineering Series,
turn to the back of this book.

Design of Semiconductor QCA Systems

Weiqiang Liu
Earl E. Swartzlander, Jr.
Máire O'Neill

ARTECH HOUSE
BOSTON | LONDON
artechhouse.com

Library of Congress Cataloging-in-Publication Data
A catalog record for this book is available from the U.S. Library of Congress.

British Library Cataloguing in Publication Data
A catalogue record for this book is available from the British Library.

Cover design by Vicki Kane

ISBN 13: 978-1-60807-687-1

685 Canton Street
Norwood, MA 02062

10 9 8 7 6 5 4 3 2 1

Contents

Part I
QCA Background

1

Introduction

Weiqiang Liu, Máire O'Neill, and Earl E. Swartzlander, Jr.

Integrated circuits have become much smaller, cheaper, and more reliable and have revolutionized the world of electronics. Currently, integrated circuits are used in almost all electronic devices and systems, many of which, such as the Internet, computers and mobile phones, have become essential parts of modern life and have changed the way we live. The first generation of integrated circuits was developed in the late 1950s. Since then, it has become possible to place more and more transistors onto a single chip with lower costs as predicted by Moore's law [1], which stated that the number of transistors that can be integrated in a chip doubles approximately every two years. As a technical roadmap, Moore's law has accurately predicted the development of the semiconductor industry for almost 50 years.

However, this scaling trend faces serious challenges due to physical limitations of current complementary metal-oxide-semiconductor (CMOS) technology such as high-power dissipation, minimum fabrication dimensions, off-state leakage, and so forth. The International Technology Roadmap for Semiconductors (ITRS) report predicts that the scaling of current CMOS technology will end by 2019 [2]. As the feature size shrinks to less than 20nm, the operation of transistors will be ruled by quantum physics, which is not considered in current technology. Therefore, there is a need for new kinds of devices that adopt quantum effects and take advantage of quantum physics.

Nanotechnologies, which offer feature sizes in the range of 1 to 100 nm provide new possibilities for computing paradigms that could be used to continue Moore's law. New devices including quantum-dot cellular automata (QCA) devices, carbon nanotubes, single electron transistors, resonant tunneling diodes, single-molecule devices, and spin transistors have been investigated [2–4]. One promising possibility based on quantum dots, that is, QCA, was introduced in 1993 by Lent et al. [5]. Although practical QCA devices are still being developed, experimental devices for semiconductor, molecular, and magnetic approaches have been explored [6–14]. Research has shown that room-temperature operation of QCA circuits is achievable [10, 15]. Currently, the fabrication of QCA circuits is very limited, and they are not commercially manufactured. Even though the practical components are not available yet, the technology needs to be fully explored to serve as a basis for future commercial fabrication.

QCA provides a revolutionary approach to computing with device-to-device interactions, which is one of the obstacles for further scaling of CMOS devices. By using device-level interaction, referred to as "processing-in-wire," rather than traditional switching, novel approaches to computation and communication are offered. As both QCA logic gates and transmission wires are composed of basic QCA cells, the computation and communication occurs at the same time. The four-phase clocking scheme applied in QCA circuits as discussed in Chapter 2 also enables deep-level pipelines. This requires the data to be stored in loop wires, which allows the new memory architecture design of "memory-in-motion" [16]. Overall, QCA has the potential advantages of high speed [17], high-density integration [18], and low-power dissipation [19].

1.1 Motivation

The design of a QCA circuit is radically different from a conventional digital design due to its unique characteristics at both the physical level and logic level. High-level designs focus on logical and algorithmic design in addition to the physical design. Even though actual QCA circuit designs need to manage considerable physical interactions that are possibly undesirable and disruptive, the algorithmic approach is also an important aspect in large systems. Research into both circuit architecture and device design is required for a profound understanding of QCA nanotechnologies. This book focuses on the logic and algorithmic level design.

As QCA devices show great potential for computation, it is necessary to study design strategies and the performance of computer arithmetic circuits based on the new characteristics of QCA. Computer arithmetic circuit design is a fundamental subject since adders, multipliers, and dividers are the most important components in an artithmatic logical unit (ALU). To exploit the characteristics of QCA technology and illustrate its potential benefits from a circuit-design perspective, a comparison with different circuits is needed, especially for large-scale designs with significant size and complexity issues. Both conventional and novel structures of arithmetic circuits should be investigated in QCA. It is also necessary to show that large-scale designs in QCA are possible.

The unique characteristics of QCA circuits present new design challenges. The four-phase clocking in QCA enables deep pipelines; however, this can lead to serious timing issues referred to as the "layout=timing" problem [20]. Although QCA logic components can be designed with QCA gates, extra delays will be introduced, which can lead to incorrect timing relationships. These timing issues present difficulties for interconnection and feedback and can significantly affect the performance of QCA circuits. Therefore, assigning correct and efficient clocking zones to circuits is a major challenge in QCA circuit design.

Design techniques that can be used to achieve optimal QCA design are not well-developed. Appropriate design methodologies are required for efficient design in QCA. Wire delays are a major factor that affect the performance of QCA circuits. For example, research has shown that although carry-lookahead adders (CLAs) are faster than ripple-carry adders (RCAs) in CMOS, an optimized layout of a RCA is faster than a CLA adder in QCA [21]. The wire delays in QCA account for most of the difference. Since an adder is a relatively simple component in digital circuit design, it is envisaged that in the case of a more complex design, wire delay will seriously affect its performance. Therefore, design methodologies are required that take into account the characteristics of QCA technology.

Although circuit designs in QCA have been extensively studied, how to properly evaluate the QCA circuits has not been carefully considered. Research [22, 23] has compared different QCA circuits by using metrics that are used in CMOS technology. Such metrics are not appropriate for QCA technology as there are fundamental differences between the two technologies. When comparing QCA circuits, some unique metrics for QCA technology must be considered. Therefore, general cost functions with appropriate cost metrics are needed in order to determine the most optimal design and to help inform QCA circuit design.

1.2 Contributions

This book provides a comprehensive introduction to semiconductor QCA circuit design. A preliminary set of important design rules for QCA is compiled in this book. Both layout and timing rules are developed. Timing rules in QCA are believed to be as important as layout rules. These rules should be followed in order to achieve robust designs [24].

The book presents an extensive study of computer arithmetic circuit designs in QCA. Several QCA adder designs are proposed, including both binary and decimal adders that are optimized for QCA technology from conventional design techniques. RCAs, pipelined CLAs and conditional sum adders (CSAs) are designed and compared up to 64-bit word sizes [21, 25–26]. Initial studies showed that the delays of the CLAs are less than that of the RCAs and CSAs when the operand size is large. Even though the complexity of the CLA design is higher than that of the RCA, overall it is the best adder design in QCA [21]. In QCA, the CSA is slower and its complexity and area are much greater than those of the CLA [21, 26]. It is also found that interconnections in QCA circuits incur significant complexity and wire delays. Based on QCA's unique characteristics, this book presents a new RCA design, the carry-flow adder (CFA). CFAs use conventional carry propagation schemes, but are optimized for layout in the QCA technology. Compared with other adder designs, the CFA shows the smallest complexity, area, and delay. Therefore, with a carefully optimized full adder design a QCA RCA is faster and smaller than a CLA in QCA [27]. Two cost-efficient binary-coded decimal (BCD) adders are also presented in this book. Both decimal adders achieve better performance in terms of latency and overall cost than previous designs [28].

Parallel multipliers including array multipliers [29], Wallace multipliers, and Dadda multipliers have been constructed and analyzed in QCA technology [30]. To facilitate modular design and to accommodate large word sizes, quasi-modular parallel multipliers are proposed in the book and compared with previous designs. It is found that the quasi-modular multipliers are much slower than expected. However, 8-bit by 8-bit tree multipliers have large areas, many cells and a longer delay than 8-bit by 8-bit quasi-modular multipliers. It is shown again that long wires significantly affect timing, with 33.8% of the latency due to the wiring. Array multipliers are the best choice for QCA implementation, as the latency is lower and the area is much less than Wallace, Dadda, and quasi-modular multipliers due to their conformability with QCA technology without the need for extra wire delay [29].

A restoring binary divider has been designed for implementation with QCA technology. It can be easily enlarged without long data connections, but it is large and slow due to the restoring algorithm. However, by using a pipelined parallel structure it has a good throughput [31]. A Goldschmidt iterative divider designed using a data tag method shows that sequential circuits in QCA can be built efficiently without state machines [32]. The proposed data tag method avoids the synchronization problems that arise with conventional state machines in QCA due to the long delays between the state machines and the units to be controlled. In the proposed architecture, it is possible to start a new division at any iteration stage of a previously issued operation. As a result, the throughput is significantly increased since multiple division computations can be performed in a time-skewed manner using one iterative divider. Using the data tag method, the Goldschmidt divider achieves a much smaller design area and has a greatly reduced latency in comparison to the restoring array divider.

A cut-set retiming design procedure is proposed to mitigate the QCA timing issues. The proposed design procedure can accommodate QCA's unique characteristics by performing delay transfer and time scaling to reallocate the existing delays so as to achieve efficient clocking zone assignment. This design procedure makes it possible to effectively design relatively complex QCA circuits that include feedback [33, 34]. A QCA Montgomery modular multiplier is designed to demonstrate that complex circuits can be designed in QCA using the proposed cut-set retiming design procedure [33]. Furthermore, an S27 benchmark circuit is designed using the cut-set retiming design procedure and compared with previous designs. The comparison shows that the cut-set retiming method achieves a more efficient design, with a reduction of 22%, 44%, and 46% in terms of cell count, area, and latency, respectively [34].

Both systolic arrays and QCA have similar characteristics, namely, ease of synchronization, deep pipelines, and local connections with simple control, all of which lead to efficient QCA circuits. Two case studies, a matrix multiplier [35] and a Galois field multiplier [24], are presented and analyzed. Based on the case studies, a systematic approach to designing QCA systolic array architectures is presented, which allows the design of efficient QCA circuits is presented. It is found that by applying a systolic array in QCA design, significant benefits can be achieved particularly with large systolic arrays, even more so than when applied in CMOS-based designs [36].

Finally, a family of new cost functions is proposed to fairly evaluate QCA circuits. Several cost metrics for QCA circuits are reviewed and studied [37]. It is found that delay, the number of majority gates, and the number and type of crossovers are important cost metrics that should be included in QCA cost functions. By using the proposed cost metrics to review a representative number of QCA adder designs, it is shown that different optimization goals lead to different "best" adders in comparison to equivalent optimization goals in CMOS.

1.3 Book Outline

Chapter 2 provides comprehensive background information on QCA technology. QCA fundamentals including QCA cells and wires, basic logic gates, and wire crossings are introduced. Four physical implementations of QCA are outlined with their advantages and disadvantages. Different clocking schemes including both typical four-phase clocking and clocking for reversible computing are presented. QCA design and simulation tools are introduced with a focus on QCADesigner. A preliminary set of important design rules for QCA and a survey on general digital design and testing by QCA are also provided.

Chapter 3 investigates the design and implementation of binary and decimal adders. Designs of QCA adders derived from conventional CMOS designs including RCAs, CLAs and CSAs are first shown. Then a new type of QCA adder, namely the CFA, is proposed based on QCA characteristics. Finally, the design of decimal adders in QCA is also discussed.

Chapter 4 explores the implementation of three types of parallel multipliers in QCA technology. Array multipliers that are well suited to QCA are constructed and formed by a regular lattice of identical functional units. Then, column compression multipliers including Wallace and Dadda multipliers are implemented with several different operand sizes. Quasi-modular parallel multipliers are also designed to facilitate the use of irregular tree structure multipliers.

Chapter 5 presents the designs of two very different types of dividers: a digit recurrent divider and a convergent divider. The restoring binary divider is implemented with controlled full subtractor (CFS) cell blocks. A convergent divider, namely the Goldschmidt divider, is implemented efficiently for QCA using a data tag method.

Chapter 6 discusses timing constraints and issues in QCA. The retiming technique is introduced and a cut-set retiming design procedure based on QCA timing constraints is proposed to resolve timing issues. Case studies including both systolic and nonsystolic architectures are presented to illustrate the proposed procedure.

Chapter 7 investigates QCA systolic array design to further explore the characteristics of QCA. Two multipliers architectures, a matrix multiplier and a Galois field multiplier, are designed and analyzed based on both a systolic array design and a single-processor design with multilayer and coplanar crossings. Then a general systolic array design methodology is proposed.

Chapter 8 proposes a family of new cost functions based on new cost metrics that could be used to evaluate QCA circuits. The evolution of cost metrics used in CMOS technology is revisited first. Then the metrics used in previous research are reviewed, and several new metrics are discussed. Based on the analysis of all available metrics, a family of cost functions is proposed. A representative number of QCA adder designs is reviewed, and general metric formulas for each adder are derived based on schematics and layouts. QCA adders are compared in terms of each important metric and then evaluated with the proposed cost functions in terms of the overall cost.

Chapter 9 provides a summary and conclusions. In addition, Chapter 9 discusses possible directions for future research.

References

[1] Moore, G., "Cramming More Components onto Integrated Circuits," *Electronics,* Vol. 38, No. 8, April 19, 1965, pp. 114–117.

[2] "International Technology Roadmap for Semiconductors (ITRS)," website, 2011, http://www.itrs. net/Links/2011ITRS/Home2011.htm.

[3] Lundstrom, M., "Is Nanoelectronics the Future of Microelectronics?" in *Proceedings of the 2002 International Symposium on Low Power Electronics and Design*, 2002, pp. 172–177.

[4] "International Technology Roadmap for Semiconductors (ITRS)," website, 2004, http://www.itrs. net/Links/2004Update/2004Update.htm.

[5] Lent, C., et al., "Quantum Cellular Automata," *Nanotechnology*, Vol. 4, 1993, pp. 49–57.

[6] Orlov, A., et al., "Realization of a Functional Cell for Quantum-dot Cellular Automata," *Science*, Vol. 277, No. 5328, 1997, pp. 928–930.

[7] Orlov, A,. et al., "Experimental Demonstration of a Binary Wire for Quantum-dot Cellular Automata," *Applied Physics Letters*, Vol. 74, No. 19, 1999, pp. 2875–2877.

[8] Amlani, I., et al., "Digital Logic Gate Using Quantum-Dot Cellular Automata," *Science*, Vol. 284, No. 5412, 1999, pp. 289–291.

[9] Lent, C., "Bypassing the Transistor Paradigm," *Science*, Vol. 288, No. 5471, 2000, pp. 1597–1599.

[10] Cowburn, R., and M. Welland, "Room Temperature Magnetic Quantum Cellular Automata," *Science*, Vol. 287, No. 5457, 2000, pp. 1466–1468.

[11] Lent, C., B. Isaksen, and M. Lieberman, "Molecular Quantum-dot Cellular Automata," *Journal of the American Chemical Society*, Vol. 125, No. 4, 2003, pp. 1056–1063.

[12] Imre, A., et al., "Majority Logic Gate for Magnetic Quantum-dot Cellular Automata," *Science*, Vol. 311, No. 5758, 2006, pp. 205–208.

[13] Haider, M., et al., "Controlled Coupling and Occupation of Silicon Atomic Quantum Dots at Room Temperature," *Physical Review Letters*, Vol. 102, No. 4, 2009, pp. 46805–46808.

[14] Eichwald, I., et al., "Nanomagnetic Logic: Error-Free, Directed Signal Transmission by an Inverter Chain," *IEEE Transactions on Magnetics*, Vol. 48, 2012, pp. 4332–4335.

[15] Wang, Y., and M. Lieberman, "Thermodynamic Behavior of Molecular-Scale Quantum-dot Cellular Automata (QCA) Wires and Logic Devices," *IEEE Transactions on Nanotechnology*, Vol. 3, 2004, pp. 368–376.

[16] Frost, S., et al., "Memory in Motion: A Study of Storage Structures in QCA," in *Proceedings of 1st Workshop on Non-Silicon Computing*, Vol. 2, 2002, pp. 30–37.

[17] Seminario, J., et al., "A Molecular Device Operating at Terahertz Frequencies: Theoretical Simulations," *IEEE Transactions on Nanotechnology*, Vol. 3, 2004, pp. 215–218.

[18] DeHon, A., and M. J. Wilson, "Nanowire-Based Sublithographic Programmable Logic Arrays," in *Proceedings of the 12th ACM/SIGDA International Symposium on Field Programmable Gate Arrays*, 2004, pp. 123–132.

[19] Timler, J., and C. Lent, "Power Gain and Dissipation in Quantum-dot Cellular Automata," *Journal of Applied Physics*, Vol. 91, 2002, pp. 823–831.

[20] Niemier, M., and P. Kogge, "Problems in Designing with QCAs: Layout=Timing," *International Journal of Circuit Theory and Applications*, Vol. 29, No. 1, 2001, pp. 49–62.

[21] Cho, H., and E. Swartzlander, Jr., "Adder Designs and Analyses for Quantum-Dot Cellular Automata," *IEEE Transactions on Nanotechnology*, Vol. 6, 2007, pp. 374–383.

[22] Gladshtein, M., "Quantum-dot Cellular Automata Serial Decimal Adder," *IEEE Transactions on Nanotechnology*, Vol. 10, 2011, pp. 1377–1382.

[23] Pudi, V. and K. Sridharan, "Efficient Design of a Hybrid Adder in Quantum-Dot Cellular Automata," *IEEE Transactions on Very Large Scale Integration Systems*, Vol. 19, 2011, pp. 1535– 1548.

[24] Liu, W., et al., "Design Rules for Quantum-dot Cellular Automata," in *Proceedings of the IEEE International Symposium on Circuits and Systems*, 2011, pp. 2361–2364.

[25] Cho, H., and E. Swartzlander, Jr., "Pipelined Carry Lookahead Adder Design in Quantum-Dot Cellular Automata," in *Conference Record of the 39th Asilomar Conference on Signals, Systems and Computers*, 2005, pp. 1191–1195.

[26] Cho, H., and E. Swartzlander, Jr., "Modular Design of Conditional Sum Adders Using Quantum-dot Cellular Automata," in *Proceedings of the 6th IEEE Conference on Nanotechnology*, Vol. 1, 2006, pp. 363–366.

[27] Cho, H., and E. Swartzlander, Jr., "Adder and Multiplier Design in Quantum-dot Cellular Automata," *IEEE Transactions on Computers*, Vol. 58, 2009, pp. 721–727.

[28] Liu, W., et al., "Cost-Efficient Decimal Adder Design in Quantum-dot Cellular Automata," in *Proceedings of the IEEE International Symposium on Circuits and Systems*, 2012, pp. 1347–1350.

[29] Kim, S., and E. Swartzlander, Jr., "Multipliers with Coplanar Crossings for Quantum-dot Cellular Automata," in *Proceedings of 10th IEEE Conference on Nanotechnology*, 2010, pp. 953–957.

[30] Kim, S., and E. Swartzlander, Jr., "Parallel Multipliers for Quantum-Dot Cellular Automata," in *Proceedings of the IEEE Nanotechnology Materials and Devices Conference*, 2009, pp. 68–72.

[31] Kim, S., and E. Swartzlander Jr., "Restoring Divider Design for Quantum-Dot Cellular Automata," in *Proceedings of the 11th IEEE Conference on Nanotechnology*, 2011, pp. 1295–1300.

[32] Kong, I., E. Swartzlander, Jr., and S. Kim, "Design of a Goldschmidt Iterative Divider for Quantum-dot Cellular Automata," in *Proceedings of the IEEE/ACM International Symposium on Nanoscale Architectures*, 2009, pp. 47–50.

[33] Liu, W., et al., "Montgomery Modular Multiplier Design in Quantum-dot Cellular Automata using Cut-Set Retiming," in *Proceedings of the 10th IEEE Conference on Nanotechnology*, 2010, pp. 205–210.

[34] Liu, W., et al., "Design of Quantum-dot Cellular Automata Circuits Using Cut-Set Retiming," *IEEE Transactions on Nanotechnology*, Vol. 10, 2011, pp. 1150–1160.

[35] Lu, L., et al., "QCA Systolic Matrix Multiplier," in *Proceedings of the IEEE Computer Society Annual Symposium on VLSI*, 2010, pp. 149–154.

[36] Lu, L., et al., "QCA Systolic Array Design," *IEEE Transactions on Computers*, Vol. 62, 2013, pp. 548–560.

[37] Liu, W., et al., "A Review of QCA Adders and Metrics," in *Conference Record of the 46th Asilomar Conference on Signals, Systems and Computers*, 2012, pp. 747–751.

2

Quantum-dot Cellular Automata

Weiqiang Liu, Máire O'Neill, and Earl E. Swartzlander, Jr.

As a replacement for CMOS technology, quantum cellular automata was proposed by Lent et al. [1] to implement classic cellular automata with quantum dots. In order to distinguish this proposal from models of cellular automata performing quantum computation, the term has been changed to quantum-dot cellular automata (QCA). QCA is a revolutionary technology that exploits the inevitable nanolevel issues to perform computing. It has potential advantages including high speed, high device density, and low-power dissipation.

Research on QCA has received considerable interest since its invention. A number of leading research groups around the world are working at both the physical and logic levels. The research group at the University of Notre Dame [2] has been leading QCA research for two decades. Most of these groups were only involved in research at the physical level before the release of the state-of-art design and simulation tool QCADesigner in 2004 [3]. Since then, QCA research at the logic level has expanded dramatically. A survey of the simulation and experimental research on the fabrication of QCA devices has been provided by Macucci [4], and an aggregation of previous work on the design and test of digital circuits in QCA has been presented by Lombardi and Huang [5]. This chapter provides background information on QCA technology and surveys previous research into QCA circuit design.

This chapter is organized as follows. Section 2.1 outlines QCA fundamentals including the general QCA model, basic gates, wires, and crossover options. Possible physical implementations of QCA circuits are overviewed in Section 2.2. The clocking schemes in QCA are introduced in Section 2.3. Section 2.4 presents the tools used for QCA design and simulation, and a survey of general digital design in QCA is presented in Section 2.5. A preliminary set of design rules for QCA is discussed in Section 2.6, and a summary of the chapter is provided in Section 2.7.

2.1 QCA Fundamentals

2.1.1 QCA Cells and Wires

A general QCA cell is a square nanostructure with four quantum dots as shown schematically in Figure 2.1(a). Dots are the places where a charge can sit. The cell is populated with two electrons that can tunnel between the four quantum dots. Tunneling action only occurs within the cell and no tunneling happens between cells. The numbering of the dots (denoted as i) in the cell goes clockwise starting from the dot on the top right: top right dot $i=1$, bottom right dot $i=2$, bottom left dot $i=3$ and top left dot $i=4$. A polarization P in a cell is defined as

$$P = \frac{(\rho_1 + \rho_3) - (\rho_2 + \rho_4)}{(\rho_1 + \rho_2 + \rho_3 + \rho_4)} \tag{2.1}$$

where ρ_i denotes the electronic charge at dot i. The polarization measures the charge configuration—that is, the extent to which the electronic charge is distributed among the four dots. Binary information is represented in QCA by

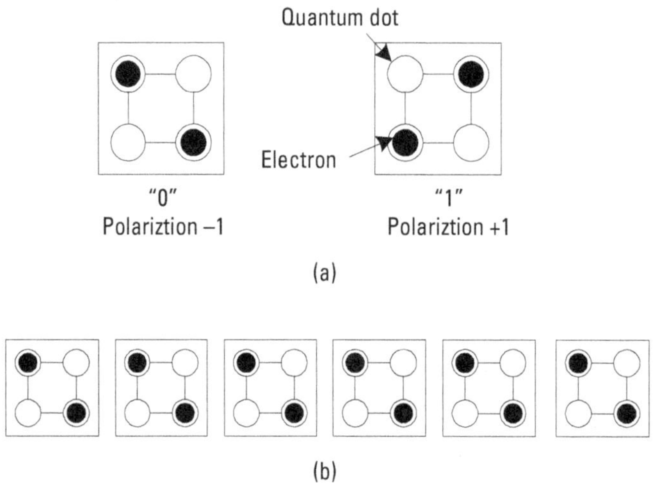

Figure 2.1 Schematics of QCA cell and wire: (a) binary QCA cells, (b) a QCA wire composed of coupled cells.

using the position of two mobile electrons in each logic cell. When the barriers between dots are low enough to free the electrons under the control of the clocking scheme, these two electrons tend to occupy antipodal sites within the cell due to Coulombic repulsion [1] as shown in Figure 2.1(a). The two charge configurations can be used to represent binary "0" and "1" with a polarization of −1 and +1, respectively. The combination of quantum confinement, Coulombic repulsion, and the discrete electronic charge produces bistable behavior.

If a cell is placed near a driver cell whose polarization is fixed, the cell will align its polarization with that of the driver cell. It has been illustrated that the cell-to-cell interaction is highly nonlinear, (i.e., even a slightly polarized input cell induces an almost fully polarized output cell) [6]. Therefore, information can be transferred by interaction between neighboring cells along a line of QCA cells. The polarization of the input can be transferred by the intercell Coulombic repulsion along the one-dimensional cell array. A QCA "wire" is a chain of cells as shown in Figure 2.1(b), where the cells are adjacent to each other rather than a physical wire. Such a wire is used as an interconnection between all kinds of logic components.

Therefore, QCA has the ability to offer "processing-in-wire" [7]. Since no electrons tunnel between cells, QCA provides a mechanism for transferring information without current flow.

2.1.2 QCA Basic Gates

Based on the mutual interaction between cells, basic logic components including an inverter and a three-input majority gate can be built in QCA. Examples of these two gates are shown in Figure 2.2. An inverter is made by positioning cells diagonally from one another to achieve the inversion functionality. A majority gate consists of five QCA cells that realize the following function:

$$M(a,b,c) = ab + bc + ac \tag{2.2}$$

Majority gates can be easily converted to AND or OR gates by using a fixed value for one of the inputs. For example, a two-input AND gate is realized by fixing one of the majority gate inputs to "0":

$$AND(a,b) = M(a,b,0) = ab \tag{2.3}$$

Similarly, an OR gate is realized by fixing one input to "1":

$$OR(a,b) = M(a,b,1) = a + b \tag{2.4}$$

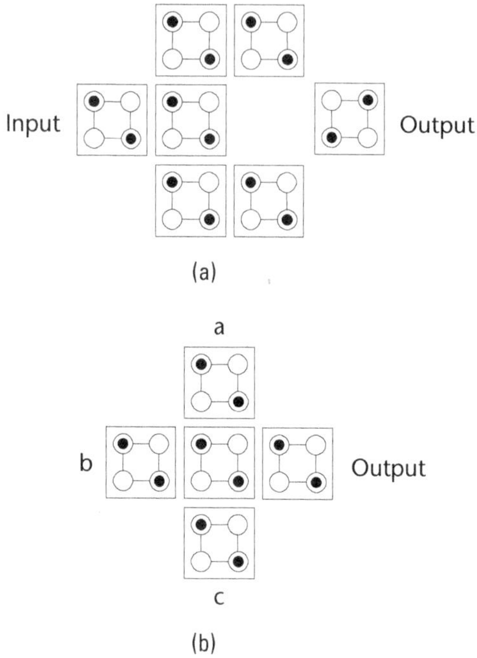

Figure 2.2 QCA basic gates: (a) inverter, (b) majority gate.

In combination with inverters, these two logic components can be used to implement any logic function.

2.1.3 QCA Wire Crossings

One of QCA's unique characteristics is the capability to create different signal wire crossings. In QCA technology, two crossover options are available: coplanar crossings and multilayer crossovers. A coplanar crossing [6] was proposed as a unique property of a QCA layout and implements crossovers by using only one layer, as shown in Figure 2.3(a), which demonstrates different forms of coplanar crossings. Half-cell displacement inverters comprised of only two regular cells, as shown in Figure 2.3(a), are used in order to propagate signals correctly. This will result in using a large number of this type of inverter in coplanar designs. A coplanar crossing uses both regular and rotated cells. The two types of cells do not interact with each other when they are properly aligned. Previous research suggests that coplanar crossings may be quite sensitive to misalignment [8] and vulnerable to noise [9, 10].

The other alternative is a multilayer crossing [11], which uses more than one layer of cells similar to the routing of metal wires in CMOS technology, as shown in Figure 2.3(b). Multilayer crossovers are expected to achieve more

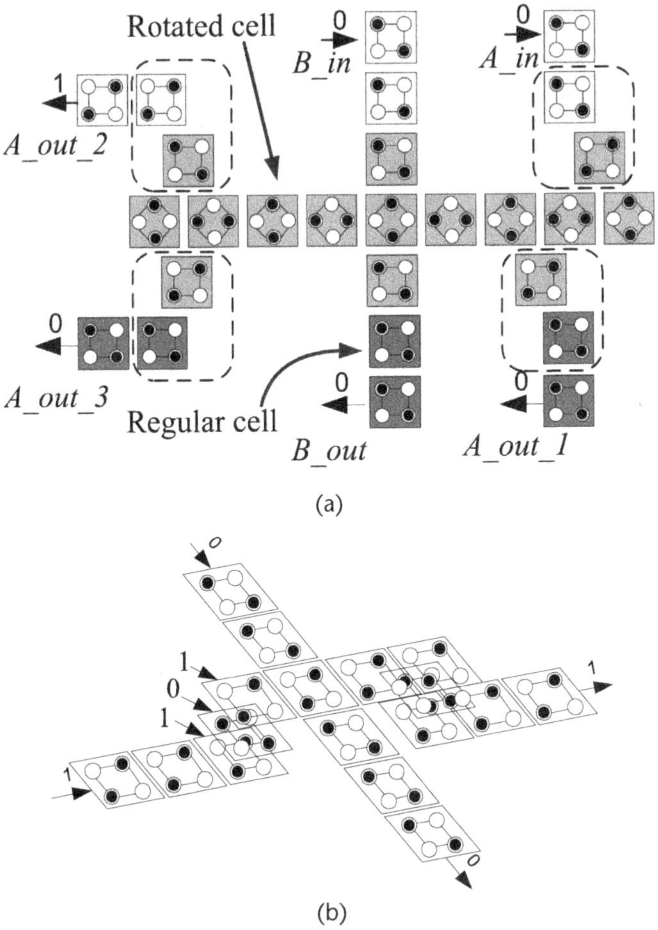

Figure 2.3 Two crossover options in QCA: (a) coplanar crossover, (b) multilayer crossover.

reliable results in simulations [12]. However, multilayer crossovers are not easy to fabricate due to the multiple layer structure, and the cost to fabricate a multilayer crossover is expected to be significantly greater than that of a coplanar crossing. The cost difference between the coplanar and multilayer crossovers affects the overall cost of a design to some extent.

2.2 Physical Implementations of QCA

To date, a number of different implementations to realize the bistable and local interaction required by the QCA paradigm have been proposed. Both electrostatic interaction-based QCA implementations (metal-dot, semiconductor, and molecular) and magnetic QCAs have been investigated. A brief overview of these four distinct classes of QCA and their advantages and disadvantages follows.

2.2.1 Metal-Island QCA

The metal-island QCA cell was implemented with relatively large metal islands (about 1 micrometer in dimension) to demonstrate the concept of QCA [13, 16]. The dots are made of aluminum with aluminum oxide tunnel junctions between them. In this metal-island QCA cell, electrons can tunnel between dots via the tunnel junctions. These two pairs of dots are coupled to each other by capacitors. Two mobile electrons in the cell tend to occupy antipodal dots due to electrostatic repulsion. Metal-island QCA components including majority gates, binary wires, memories and clocked multistage shift registers have been fabricated [16–19]. The operating temperature for metal-island QCA is extremely low, in the range of milli-Kelvin, to achieve the appropriate electron filling. This prevents the construction of complex QCA circuits running at room temperature. Therefore, the metal-island implementation is not currently seen to be a practical approach for future QCA systems.

2.2.2 Semiconductor QCA

A semiconductor QCA cell is composed of four quantum dots manufactured from standard semiconductive materials [20–22]. A device was fabricated in [23] using a GaAs/AlGaAs heterostructure with a high-mobility two-dimensional electron gas below the surface. Four dots are defined by means of metallic surface gates. The cell consists of two double Quantum-Dot systems (half cells). Half cells are capacitively coupled. The charge position is used to represent binary information and the Quantum-Dot interactions are dependent on electrostatic coupling [24]. A semiconductor implementation promises the possibility of fabricating QCA devices with the advanced fabrication processes used for existing CMOS technology. However, current semiconductor processes cannot provide mass production with the ultrasmall feature sizes required by QCA technology. To date, most QCA device prototypes have been demonstrated with semiconductor implementations. Hence, the research presented in this book is conducted based on semiconductor QCA. However, the conclusions drawn from the research based on semiconductor implementation are applicable to other implementation types.

2.2.3 Molecular QCA

A molecular QCA cell [25–28] is built out of a single molecule, in which charge is localized on specific sites and can tunnel between those sites. In the molecule shown in [29], the free electrons are induced to switch between four ferrocene groups that act as quantum dots due to electrostatic interactions, and a cobalt group in the center of the square provides a bridging ligand that acts as a tunneling path. The molecules are expected to be as small as 1 nm or even smaller, which promises room-temperature operation, ultrahigh density and high speed in the terahertz range. Room-temperature operation of a molecular QCA cell has been experimentally confirmed [30]. The difficulty in realizing molecular QCA is due to the high-resolution synthesis methods and positioning of molecule devices. New construction methods for molecular QCA, including self-assembly on DNA rafts, are under investigation [31]. However, it is still very difficult to fabricate molecular QCA systems with current technologies.

2.2.4 Magnetic QCA

A magnetic QCA cell is an elongated nanomagnet with a length of around 100 nm and a thickness of 10 nm [32–34]. The shape of the nanomagnet varies for different schemes. The binary information in magnetic QCA cells is based on their single domain magnetic dipole moments. The usage of magnetic interaction inherently minimizes the energy. Although its operating frequency is rather low (around 100 MHz), it has the advantage of room-temperature operation, extremely low power dissipation and high thermal robustness. A three-input majority gate in magnetic QCA has been fabricated [34]. The first large-scale QCA systems appear to be possible with a magnetic QCA circuit, which has fewer challenges during the manufacturing process compared with other implementations [33].

2.3 Clocking Schemes

Functional QCA circuits need to be clocked in order to operate correctly. The transitions of QCA states occur under the control of potential barriers between the quantum dots in QCA cells, and it is the QCA clock that lowers and raises the tunneling barriers. Clocking in QCA not only controls data flow but also serves as the power supply [35].

2.3.1 Typical Four-Phase Clocking

Quasi-adiabatic four-phase clocking is typically used in QCA circuits offering deep pipelines. Quasi-adiabatic switching ensures the system is always in its instantaneous ground state, which significantly reduces metastability issues [36]. In quasi-adiabatic switching, QCA cells are timed in four successive clocking zones. A calculation is performed in one clocking zone. Its state is then frozen and used as the input to a successor zone. During the calculation, the successor zone is kept in an unpolarized state so that it has no influence on the predecessor zone.

The four clocking phases exist in each QCA clocking zone and there is a 90° phase shift from one clocking zone to the next, which is shown in Figure 2.4(a). The clock signals of QCA circuits are generated by applying an electric field to the QCA cells to modulate the tunneling barrier between dots (i.e., the interdot barrier). The electric field can be generated by CMOS circuits or carbon nanotubes [37]. The cells are in the HOLD phase when the interdot barrier is high and in the RELAX phase when the interdot barrier is low. When the inter-dot barrier changes from low to high or high to low, the cell is in the SWITCH or the RELEASE phase, respectively. The transition of information occurs during the SWITCH phase. A cell is latched while it is in the HOLD phase. Since the cells in one clocking zone become latched and remain in this state until the cells are latched in the next clocking zone, a clocked QCA "wire" can be treated as a chain of D-latches. The smallest unit of delay in QCA is a clocking zone delay (D^{-1}), which is a quarter of a clock cycle delay(Z^{-1}). As shown in Figure 2.4(b), the following relationship holds:

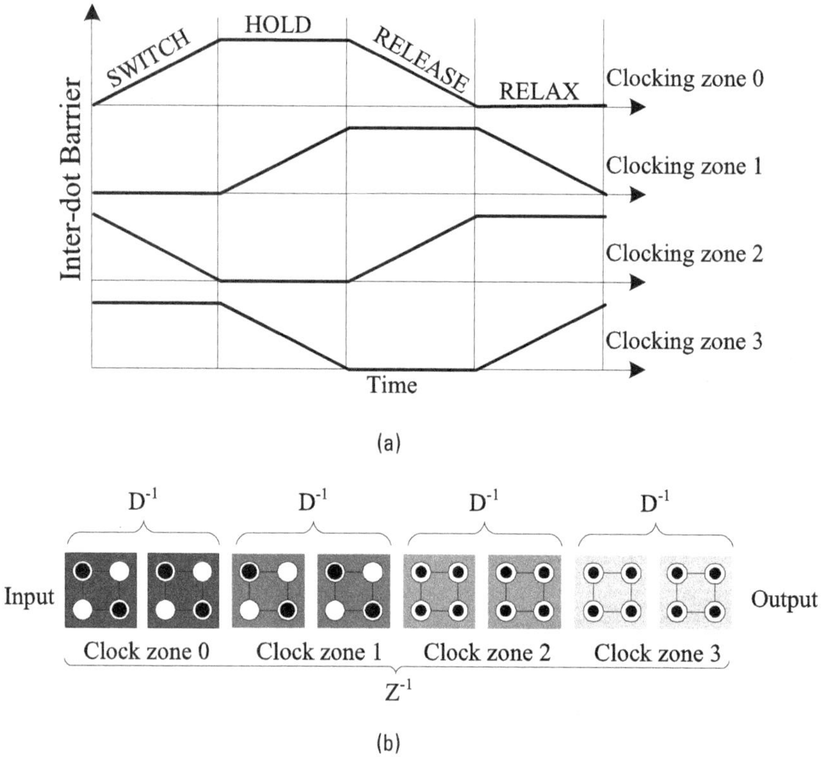

Figure 2.4 Typical QCA clocking scheme: (a) clock signals in four clocking zones, and (b) a clocked QCA wire.

$$Z^{-1} = D^{-4} \tag{2.5}$$

Although conventional logic functions can be easily mapped to majority logic, the unique clocking scheme in QCA makes it difficult to translate a CMOS architecture directly into its QCA counterpart. Although other clocking schemes have been proposed for magnetic QCA [38], in this book four-phase clocking is assumed.

2.3.2 Clocking Floorplans

Two types of clocking floorplans [39] can be used in QCA circuit implementations, namely columnar regions [Figure 2.5(a)] and zone regions [Figure 2.5(b)]. The columnar approach is assumed to be more practical in physical implementation. However, it has difficulty in realizing short feedback loops and high circuit densities. On the other hand, the zone approach can achieve these aspects. Smaller clocking zones also increase the robustness of a QCA circuit in terms of thermal effects [40]. The size of a zone is a trade-off between imple-

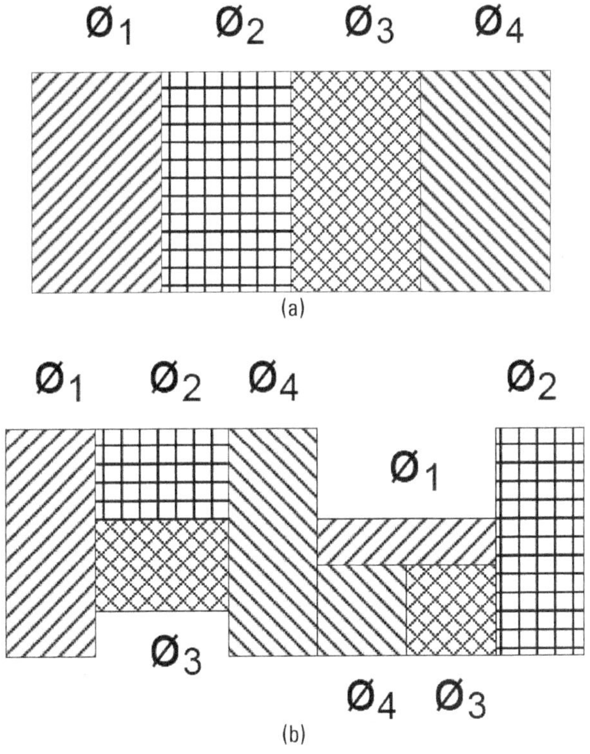

Figure 2.5 QCA clocking floorplans: (a) columnar region, and (b) zone region.

mentation difficulty and circuit efficiency. Smaller zones are more difficult to implement but can be more area-efficient. The floorplan of the QCA timing zones has a significant impact on the actual layout of a QCA circuit. This is sometimes referred to as the "layout = timing" problem [41]. In this book, the main focus is on the functionality of circuits. Therefore, the clocking floorplans of the proposed QCA architectures are designed using small zones.

2.3.3 Clocking for Reversible Computing

An alternative clocking scheme, namely Bennett clocking, was proposed as a practical means to perform reversible computing with lower power dissipation in QCA circuits [42]. The principle of Bennett clocking is to keep copies of the bit information by echoing inputs to outputs [43]. An example of Bennett clocking waveforms is shown in Figure 2.6. The timing of the Bennett clock is altered in order to keep the bit information in place until a computational block is finished. Then the information is erased during the reverse order of computation. To implement Bennett clocking in QCA, only the timing of the clocking is required to be altered without changing the circuit itself. Bennett clocking can reduce the power dissipation to even less than $k_B T\ln(2)$ [42]. However, this is achieved at the cost of speed. A general floorplan was also proposed for reversible QCA circuits [44].

2.4 Design and Simulation Tools

Several design and simulation tools for QCA circuits have been developed by using approximations with low computational complexity, which can be used for relatively large scale circuit layout and simulation. These tools include MA-

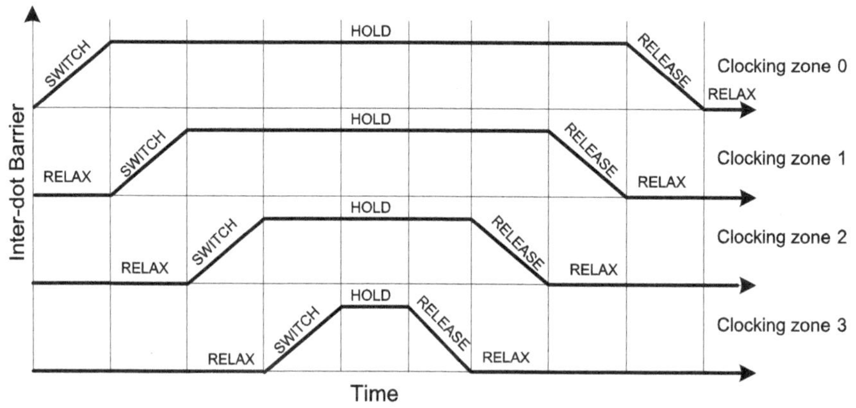

Figure 2.6 Bennentt clocking waveforms.

QUINAS [45, 46], QBART [47] and QCADesigner [3]. A SPICE macro model for QCA has also been proposed and experimentally verified [48], in addition to the hardware description language (HDL)–based design tool HDLQ, for verifying the logic behavior of QCA circuits [49]. More recently, a number of add-on features for the QCADesigner tool have been developed. These include an automatic layout generator for combinational circuits in QCADesigner [50] and a simulator for power dissipation and error estimation known as QCAPro [51].

Of these simulation tools, QCADesigner is the state-of-the-art and the most widely used QCA simulation tool. QCAPro is the first simulator for estimating both the polarization error and power dissipation in QCA circuits. These two design tools are introduced here.

2.4.1 QCADesigner

QCADesigner [3] is the most popular simulation tool for semiconductor QCA circuit design. It allows users to quickly layout a QCA design and determine its functionality in a reasonable time frame. QCADesigner supports both coplanar and multilayer crossings. The design flow of QCADesigner is shown in Figure 2.7. In the current version, QCADesigner ver 2.0.3 [52], there are two simulation engines: the bistable engine and the coherence vector engine.

2.4.1.1 Bistable Simulation Engine

The Bistable engine is implemented with each cell modeled as a simple two-state system. The following Hamiltonian matrix can be used to describe the two-state system:

$$H_i = \begin{bmatrix} -\frac{1}{2} P_j E_{i,j}^k & -\gamma \\ -\gamma & \frac{1}{2} P_j E_{i,j}^k \end{bmatrix} \tag{2.6}$$

where, P_j is the polarization of cell j. E^k is the kink energy between cell I and j. This kink energy is associated with the energy cost of two cells having opposite polarization γ is the tunneling energy of electrons within the cell, which is controlled by the clock.

The engine uses the intercellular Hartree approximation (ICHA) [1] [6] [53] to solve a quantum mechanical system by treating each individual cell quantum-mechanically and coupling neighboring cells based on the Coulombic interaction between cells. ICHA is a very important method to determine the stable state of multicell QCA systems, as it is the foundation of almost all QCA work that has been done so far [54].

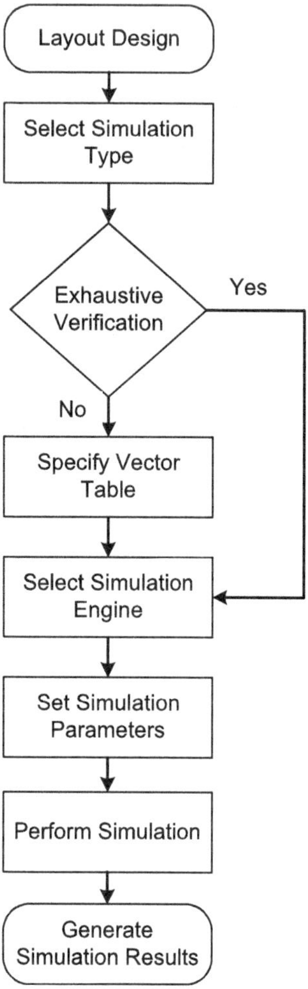

Figure 2.7 The design flow of QCADesigner.

In the ICHA method, the sum of the Hamiltonian is over all cells within an effective radius of cell *i*. Only the effects of cells that fall into a circle defined by the radius-of-effect, *R* (in nanometers), are considered for each cell. *R* is an important parameter that can be set before simulation. It is assumed that the circuits remain very close to the ground state during switching which is quasi-adiabatic. Therefore, the stationary state of each cell can be calculated by solving the time-independent Schrödinger equation [3] as follows:

$$H_i \Psi_i = E_i \Psi_i \tag{2.7}$$

where, H_i is the Hamiltonian described by (2.6). ψ_i is the state vector of the cell, and E_i is the energy associated with the state. To verify the logical functionality of a design, this eigenvalue problem reduces to performing the following calculation:

$$P_i = \frac{\frac{E_{i,j}^k}{2\gamma}\sum_j P_j}{\sqrt{1+\left(\frac{E_{i,j}^k}{2\gamma}\sum_j P_j\right)}} \tag{2.8}$$

The bistable engine iteratively computes the polarization of each cell in the circuit until the whole circuit converges to within a preset tolerance.

Full quantum mechanical calculation including quantum correlation effects within and between cells is computationally intractable for large QCA systems due to the exponential growth of the size of the Hamiltonian. By using the ICHA method, the computation only increases linearly with the number of QCA cells. Therefore, the bistable engine using ICHA is able to simulate large QCA circuits in a very efficient manner. Recent work [54] has shown that the ICHA method is generally valid and sufficient for verifying the functionality of a QCA circuit. However, the price of the quick computation is inaccuracy of the QCA dynamics due to ignoring the intercellular entanglement [55], which sometimes causes incorrect logical outputs of the QCA circuits. In order to overcome the shortcomings of the ICHA method, modifications to the circuits are usually required to make the designs robust. It should be noted that these modifications may not be necessary in practice. It is important to emphasize that the designs in this book are good starting points, but they will need refinement before actual circuits are fabricated.

2.4.1.2 Coherence Vector Simulation Engine

The coherence vector engine is a more accurate engine that provides dynamic simulation. It is based on the density matrix approach and is commonly used in simulation of QCA dynamics. The cells in the engine are modeled similarly to those in the bistable engine. However, it includes the power dissipative effects and performs a time-dependent simulation. In the density matrix approach, the coherence vector λ is a vector representation of the density matrix of a cell. The motion for the coherence vector including dissipative effects can be described as follows [35]:

$$\frac{\partial \vec{\lambda}}{\partial t} = \vec{\Gamma} \times \vec{\lambda} - \frac{1}{\tau}\left(\vec{\lambda} - \vec{\lambda}_{ss}\right) \tag{2.9}$$

where, $\vec{\Gamma}$ an energy vector representing the energy environment of the cell,

$$\vec{\Gamma} = \frac{1}{\hbar}\left[-2\gamma, 0, \sum_{j \in S} E_{i,j}^{k} P_j\right] \tag{2.10}$$

is the relaxation time that is implementation dependent and $\vec{\lambda}_{ss}$ is the steady state coherence vector,

$$\vec{\lambda}_{ss} = -\frac{\vec{\Gamma}}{\left|\vec{\Gamma}\right|}\tanh\left(\frac{\hbar\left|\vec{\Gamma}\right|}{2k_B T}\right) \tag{2.11}$$

where, $\hbar$ is the reduced Planck constant, T is the temperature in Kelvin and k_B is the Boltzmann's constant. The coherence vector for each cell is calculated by Equation (2.9) using an explicit time marching algorithm. For each time step the Γ and λ_{ss} for each cell is evaluated and then the coherence vector for each cell is stepped forward in time. For more details on the dynamics of QCA systems using the coherence vector formalism, refer to [35, 56].

It is computational expensive to determine the steadily state density matrix [55]. Meanwhile, simulation using either a bistable engine or a coherence vector engine provides the same result for semiconductor QCA circuits in most cases. Therefore, a bistable engine is usually used in the simulation of QCA circuits.

2.4.1.3 Simulation Parameters

In the current QCADesigner version 2.0.3, the size of the basic quantum cell was set at 18 nm by 18 nm with 5 nm diameter quantum dots. The center-to-center distance is set at 20 nm for adjacent cells. Research with silicon atom dangling bonds [30] shows the potential to operate at room temperature with cell sizes on the order of 2 nm × 2 nm, which will reduce the area by two orders of magnitude. The larger size was used in this research to maintain consistency with other recent QCA designs.

There are several parameters that can be set by designers in the bistable engine. The total simulation is divided by the number of samples. For each sample, the simulation engine looks at each cell and calculates its polarization based on the polarizations of its effective neighbors that are determined by the radius of effect. The number should not be chosen too small as there will be insufficient samples to get the correct results. However, a larger number results in a longer simulation time. If the simulation results should be something other than that they are, a larger number should be used [52]. The radius of effect determines how far each cell will look to find its neighbors'. The least radius of effect should include the next-to-nearest neighboring cells. For multilayer crossings, the radius of effect should be greater than that of the layer separation. In this book, the default value (i.e., 65 nm) is used. Based on the above reasons, different numbers of samples are selected for each simulation in the following chapters. All the other parameters that are the defaults for the bistable approximation are listed as follows:

- Convergence tolerance = 0.001;
- Radius of effect = 65 nm;
- Relative permittivity = 12.90;
- Clock high = $9.80e$ -22J;
- Clock low = $3.80e$ -23J;
- Clock amplitude factor = 2.00;
- Layer separation = 11.50 nm;
- Maximum iterations per sample =100.

2.4.2 QCAPro

QCAPro is a probabilistic modeling tool that can be utilized to estimate the polarization error and power dissipation under abrupt switching in QCA circuits. It is a graphic user interface (GUI)-based tool built on the Bayesian network [57, 58]. The tool can estimate erroneous cells in large QCA circuit designs by fast approximation.

It also estimates switching power loss in QCA circuits using the upper bound power model [59]. Several parameters can be used to analyze and optimize QCA designs. Users can set values for temperature and tunneling energy. QCAPro estimates the upper bound of power dissipation as a function of cell polarization, clock energy, and quantum relaxation time. The input required for the current version, QCAPro 1.0 [51], is the layout file generated by QCADesigner. It can provide the average, maximum, and minimum power consumption of a QCA circuit during input switching. The design flow of QCAPro is shown in Figure 2.8.

2.5 Research Into QCA Digital Design

Although the implementation of QCA technology is still at an early stage, researching high-level logic design is as important as physical design and can help guide its development. QCA technology not only provides a fundamentally novel physical structure, but also offers a new kind of computing architecture for digital design. The special "processing-in-wire" and "memory-in-motion"

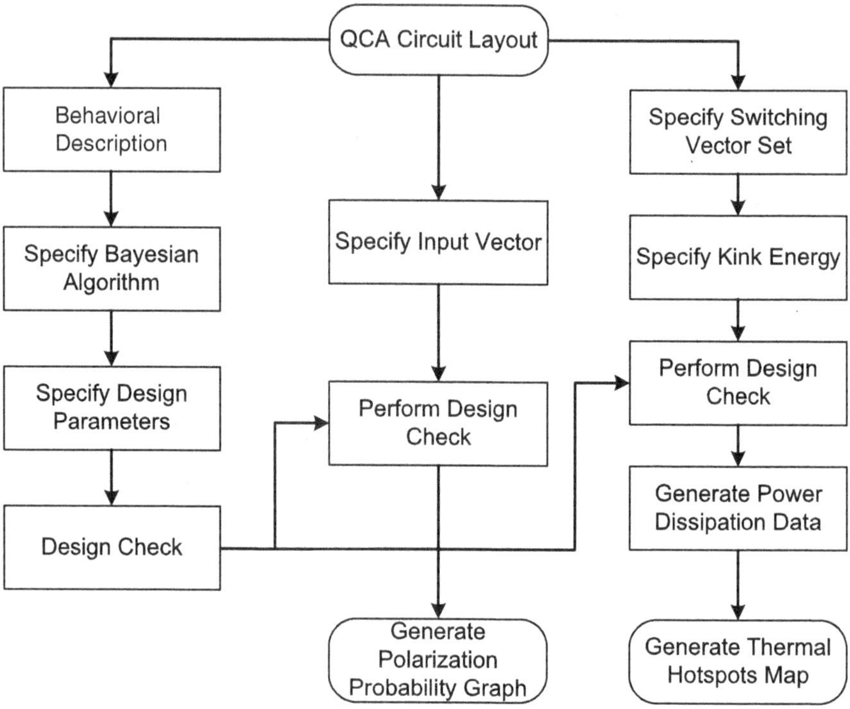

Figure 2.8 The design flow of the QCAPro tool. (*From* [51]. © 2011 IEEE.)

features [7] require the development of novel circuit architectures and new design methods that are different from traditional CMOS technology. The unique characteristics of QCA technology also present new challenges for design and testing.

2.5.1 Computer Arithmetic Circuits

Among the circuits designed in QCA, adders and multipliers have received considerable interest due to their importance in computing systems [60–62]. The first QCA circuit proposed was a 1-bit full adder [6] using five majority gates. The design was further optimized by Wang et al. [63], to use only three majority gates and three crossovers, thus significantly reducing the complexity of the adder. Wang's adder was later revised for multilayer implementation [64]. The Hänninen adder [65, 66] uses the same addition algorithm as Wang's adder, but with an optimized layout. The clocking zones were rearranged to achieve a more robust design. A CFA was proposed by Cho and Swartzlander [67], and is a layout optimized multilayer full adder. This QCA CFA consumes only one clocking zone delay per bit, which significantly reduces the overall delay in large adders. Cho and Swartzlander also designed and analyzed a CLA [68] and

CSAs [69, 70]. More recently, a family of prefix adders (which are variations of CLAs) including Kogge-Stone, Brent-Kung, Ladner-Fisher and Han-Carlson adders, were designed in QCA by reducing the carry computation to a prefix computation [71, 72]. By using a new majority logic reduction technique, the prefix adders achieve the best performance to date in terms of delay, especially for large adders.

Binary multiplier designs based on the direct paper-and-pencil algorithm have also been extensively studied in QCA. The first QCA multiplier proposed was a bit-serial multiplier with one operand in bit-serial format and the other in parallel format [60]. The design was further optimized to perform more robustly by Hänninen and Takala [73]. Cho and Swartzlander designed a serial parallel multiplier based on filter networks using a bit-serial systolic array structure [74] [67]. Fast multipliers have also been proposed by using Wallace and Dadda approaches to reduce the propagation delays [75]. Array multipliers were studied in QCA in which both operands arrive in parallel [76, 77]. A radix-4 recoded multiplier was also designed using modified Booth recoding and carry-save addition to achieve stall-free pipeline operation [78].

Other research into QCA computer arithmetic circuits has included the design of an iterative Goldschmidt divider [79] and a restoring divider [80]. A novel QCA matrix multiplier was recently proposed [81] based on majority gates, data flow using quasi-adiabatic switching, an OR loop memory, and a tristate buffer. Systolic matrix multipliers of varying size and dimension have been designed and analysed [82, 83]. Galois Field multipliers [84, 83] and Montgomery multipliers [85, 86] for cryptographic algorithms have also been designed. Decimal arithmetic for specific applications has also been studied in QCA [87–90].

2.5.2 Combinational Circuits

Combinational circuits based on majority gates and inverters have been considered in QCA. A universal gate, the and-or-inverter (AOI) gate [91], was proposed to efficiently implement elementary gates. Two-level logic functions can be easily implemented by a single AOI gate. Various multiplexer architectures [92–94] have been designed for more complex circuits. A 4-bit barrel shifter comprising a shifting unit, decoder and serial OR array was designed by Vetteth et al. [95] and Huang et al. [96] designed a decoder and parity checker using a modular tile-based method.

2.5.3 Latches and Sequential Circuits

The basic elements of sequential circuits including latches and counters have been investigated. R-S latch [97, 98], D latch [98–101] and J-K latch [102, 103] designs in QCA have been studied. Logically reversible latches were designed based on the basic reversible Toffoli and Fredkin gates [104]. However, how to design the latches in QCA is still an open issue. A state machine design in QCA is also a challenge. As a starting point, counter designs have been proposed such as a Gray code counter [98], a ring counter [99], and a synchronous counter [103]. A traffic light controller and an ISCAS89 S27 benchmark were also designed in QCA using a stretching algorithm for delay matching [98]. A data tag method was proposed [79] as an alternative way to control the various elements of the machine.

2.5.4 Memory Design

Unlike CMOS technology, there is no QCA equivalent of a capacitor to keep the state. The states must be kept in a ring of QCA cell arrays through a loop of clocking zones, which is sometimes referred to as "memory-in-motion" [7]. QCA memory structures can be categorized as two types: loop-based and line-based memory. The loop-based memory cell [7, 105–108] stores data with a closed QCA wire loop which is partitioned into four clocking zones. As a result, a large number of clocking zone delays are introduced. In a line-based memory cell [109–111], information is stored through a QCA line with a revised clocking scheme to achieve higher memory density.

2.5.5 General and Specific Processors

A microprocessor named "Simple 12" was designed in QCA [41, 47, 112]. The design was further improved to achieve robust operation in the presence of sneak noise paths [9]. A 4-bit processor design based on an accumulator architecture shows that QCA technology could potentially be applied in future computers [8, 12, 113]. Programmable logic has also been studied to implement universal logic in QCA [114–117].

Special purpose processors have also been investigated to perform signal processing tasks. For example, nonlinear filters based on QCA arrays were proposed for signal processing algorithms [118], and filters and an array processor were proposed for image processing in QCA [119, 120]. Cryptographic processors for block ciphers and stream ciphers have also been designed [121, 122].

2.5.6 Design Methods and Design Automation

General design methods to achieve large-scale modular and efficient QCA circuits are important. A specific arrangement of clocking—namely, trapezoidal clocking—was proposed to reduce the design area [41]. This was also considered to be a possible method to implement feedback paths. Tile-based modular design was studied to implement versatile logic [96] and based on the conventional concept of a flip-flop, and a stretching algorithm was proposed [98] for assigning clocking zones to QCA sequential circuits by matching delays. The delay-matching design method ensures that all paths from the outputs to the inputs of flip-flops have the same delays. However, many unnecessary delays are introduced due to the strict matching strategy resulting in an expansion of the overall number of cells and circuit size.

Two-dimension clocking was proposed [97] to reduce the longest line length in each clocking zone. A globally asynchronous, locally synchronous (GALS) method was also proposed to reduce the "layout=timing" dependency [123–125]. A cut-set retiming procedure as described in Chapter 6, was proposed to resolve the timing issues in QCA [85, 86] and general systematic approaches for the design of systolic array architectures, as outlined in Chapter 7, have been studied [82, 83].

Research into design automation, including logic synthesis, placing, routing and layout, is required to deal with the design challenges in QCA systems. As majority gates are the logic primitive in QCA, logic synthesis based on majority logic has been extensively studied [126–129]. A tool referred to as the majority logic synthesizer (MALS) was developed for general multilevel majority/minority network synthesis [130, 131]. An algorithm targeting logic-level abstraction for area minimization was also proposed for QCA circuits [132] and automatic partitioning and placement for the generation of QCA layouts have been investigated [133, 134].

2.5.7 Testing, Defects and Faults

The testing of defects and faults is critical in nanoscale integration such as QCA circuits. Although it is difficult to address the testing of QCA devices at this early stage, this aspect has been investigated by theoretical analysis and simulation. Fault models and prototype tools have been developed [135, 136]. Unique testing features and the defect characteristics have also been identified and studied [137–139]. The scaling of QCA basic gates in the presence of defects from process variations has been evaluated [140] and defect tolerance properties in tile-based QCA design studied [141]. The reliability dependence on the failure rates of macro components for arithmetic circuits has been investigated [142, 143]. Other research in this area has included the study of defect characterization and tolerance in sequential circuits [144, 145], an analysis of the displacement tolerance of QCA interconnects [146] and a discussion on the defects and faults in QCA programmable logic [147]. An information-theoretic method was also proposed to analyze the defect tolerance of QCA circuits [148]. The testing and fault tolerance of reversible QCA circuits has also been investigated [149–151].

2.6 Basic Design Rules[1]

When mapping a digital design to majority logic-based QCA circuits, knowledge of the layout and timing constraints is necessary. The objective of defining design rules is to simplify mapping from a circuit schematic to an actual layout implementation. Design rules have played an important role in the development of CMOS technology. A CMOS design rule set specifies certain geometric and connectivity restrictions to ensure sufficient margins that take into account variability in the semiconductor manufacturing processes to ensure that circuits function correctly. Limited research has been conducted into defining design rules for QCA circuits [152, 153] and guidelines for achieving robust QCA designs [10]. Developing design rules for QCA technology will help designers understand QCA features and how to efficiently achieve correct functionality and reliability in QCA circuits. This will ultimately help to promote the development of practical QCA systems.

Due to the unique clocking scheme used in QCA, there is a critical relationship between the layout and timing, referred to as "layout=timing" [41]. Consequently, the timing rules are as important as the layout in QCA. The careful placement of the cells to satisfy both types of rules can produce a more reliable design. Based on the research conducted into QCA circuit designs, a set of basic QCA design rules has been compiled [84]. Both the layout and timing rules are discussed in this section. Note that most rules are based on QCADesinger using the ICHA.

2.6.1 Layout Rules

A design rule set specifies certain geometric and connectivity restrictions to ensure that circuit components operate correctly. Following this concept, some layout design rules for QCA are described with regard to the following:

1. The maximum number of cells in a clocking zone;
2. The minimum number of cells in a clocking zone;
3. The minimum wire spacing for signal separation.

2.6.1.1 The Maximum Number of Cells in a Clocking Zone

QCA computation is achieved by relaxing the physical array to its ground state. Computing with the ground state has the undesired accompanying effect of being temperature sensitive. Thermal fluctuations may excite the QCA array above its ground state, which may produce an incorrect output. A complete analysis of thermodynamic effects has been conducted by Lent et al. [40]. As more cells are placed in a single clocking zone, more errors may occur. The

1. Section 2.6 is based on [84].

limitation on the number of QCA cells to avoid undesired kink effects is given by [40]:

$$N \le e^{\frac{E_k}{k_B T}} \tag{2.12}$$

where, N is the number of cells in the array, E_k is the kink energy between two cells, k_B the Boltzmann constant, and T is the operating temperature. The maximum operating temperature is affected by the QCA cell size. The different forms of QCA (semiconductor, magnetic, and molecular) have different kink energies, which will result in different wire length constraints [154].

Long QCA wires also result in an increased delay in signal propagation and switching, which can significantly reduce the overall operating speed. Therefore, the clock rate can be improved if a small number of cells are set into a single clocking zone. Long QCA wires should be partitioned into different clocking zones to ensure correct functionality.

2.6.1.2 The Minimum Number of Cells in a Clocking Zone

A clocking zone can contain only a single QCA cell. However, the waveform of a one-cell clocking zone can become distorted, and cascading of this kind of clocking zone could lead to incorrect results [9]. In the simulation of a wire in QCADesigner (in a range of radius of effect from 21 to 80 nm) as shown in Figure 2.9, the signals in clocking zone 1 are distorted and become worse in clocking zone 2 and finally lead to a wrong output in clocking zone 3. To achieve a reliable result, it is suggested that in most cases there should be at least two cells in each clocking zone. For a long QCA wire, the cells should be put into different clocking zones and divided evenly to avoid the effects of a one-cell clocking zone and to ensure robust signal transmission.

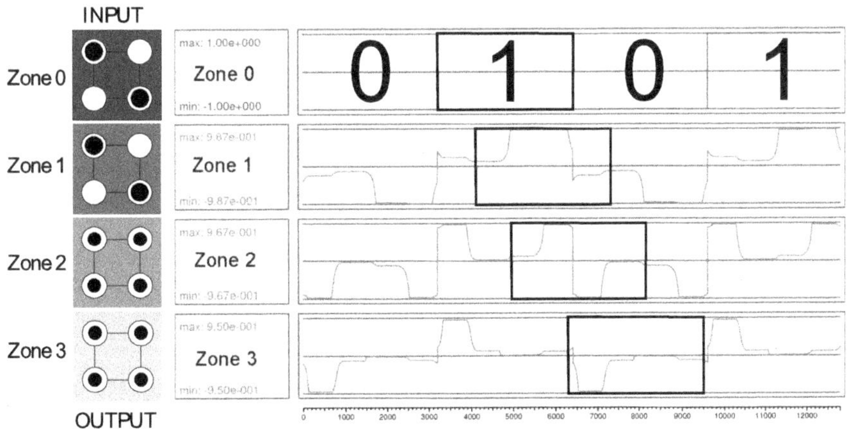

Figure 2.9 Distorted waveforms from one-cell clocking zones. (© 2011 IEEE. *From* [84].)

2.6.1.3 The Minimum Wire Spacing for Signal Separation

QCA cells interact through a quadrupole-quadrupole interaction, which decays inversely by a power of five of the distance between cells. Referring to Figure 2.10(a), the relationship between the cell position and the kink energy is as follows [152]:

$$E_k(r,\theta) \propto r^{-5}\cos(4\theta) \tag{2.13}$$

where, r is the distance between two cell centers.

If two QCA cells are aligned properly with a center-to-center distance of one cell size as shown in Figure 2.10(b), the kink energy between them is proportional to r_1^{-5} For two cells with a center-to-center distance of two cells, the kink energy is proportional to $\frac{1}{32}r_1^{-5}$ which is 32 times smaller. Therefore, the kink energy will decay rapidly with distance, and the effective neighborhood of interacting cells can be reduced. In the QCADesigner tool, the effective neighborhood of interacting cells is determined by the radius of effect. When the next-to-nearest neighbors are included in this radius, a space of one QCA

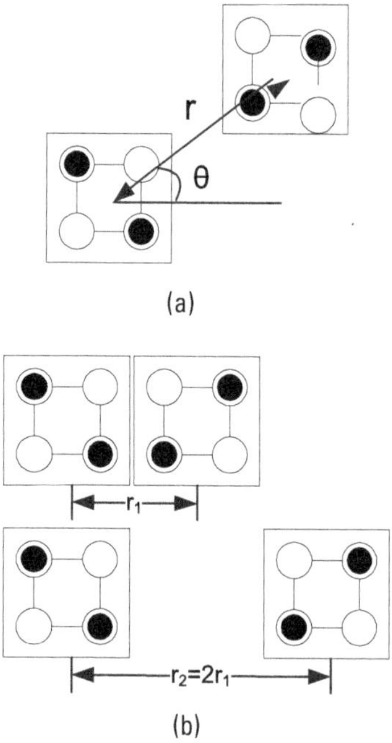

Figure 2.10 Relationship between QCA cell position and kink energy: (a) general interaction between two cells, (b) two cells with a center-to-center distance of one cell. (© 2011 IEEE. *From* [84].)

cell size is sufficient separation between two wires carrying different signals. However, for a larger radius of effect, more space is required between QCA signal wires.

2.6.2 Timing Rules

A successful QCA layout is largely determined by an appropriate clocking zone assignment due to the unique "layout = timing" aspect of QCA. Therefore, when considering QCA design rules, the timing rules are as important as the layout rules and include the following:

1. The logic component timing rule;
2. Majority logic reduction to mitigate timing constraints;
3. The clocking zone assignment rule.

2.6.2.1 Logic Component Timing Rule

The timing constraint on a QCA majority gate is that all three inputs are expected to reach the device cell (central cell) at the same time in order to have fair voting. If all three input wires are equally long, the device cell can be within the same clocking zone as the inputs. However, in practice, the length of input wires is usually different. Therefore, these three inputs should be designed within the same clocking zone i, and the majority gate as well as its output should be in the successive clocking zone $[(i+1)\ mod 4]$. As a result, at least one clocking zone delay (denoted as D^{-1}) is required in a majority gate. A robust majority gate design is shown in Figure 2.11. Thereupon, the minimum delay in its derivatives, (that is the OR and AND gate) is also one clocking zone delay (D^{-1}). However, the QCA inverter has only one input that does not need to be synchronized. Therefore, the QCA inverter does not require extra clocking zone

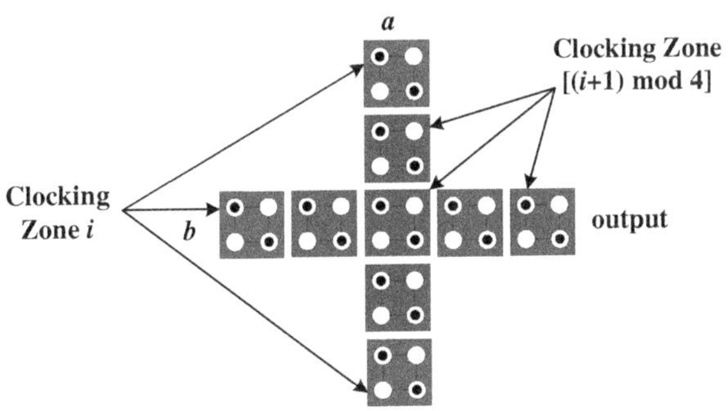

Figure 2.11 A robust QCA majority gate design. (© 2011 IEEE. *From* [86].)

delays. It has been found this timing rule comes from the simulations using the ICHA method [155]. However, this timing rule should be followed to make a robust design in QCADesigner.

2.6.2.2 Majority Logic Reduction

The logic primitive used in QCA is the majority gate. Although conventional AND and OR gates can be derived from the majority gate, it is costly in terms of cell-count to design QCA circuits by directly mapping from the equivalent CMOS design. Majority logic based reduction methods [126, 129] can significantly mitigate the timing constraints of QCA circuits and reduce the circuit complexity. A QCA design should be optimized using majority logic reduction before translating it into the layout.

2.6.2.3 Clocking Zone Assignment Rule

In QCA circuits, even combinational logic, as defined in CMOS, should be synchronized. It is easy to assign clocking zones to a signal-forward architecture in QCA, as this only requires adding delays, but for conventional sequential circuit architectures, especially those with feedback, the clocking zone assignment may be very difficult. The QCA cut-set retiming procedure as proposed by Liu et al. [86] and summarized in Chapter 6 can be used to assign correct clocking zones in complex architectures with feedback.

2.7 Summary

This chapter provides comprehensive background information on QCA technology. The general QCA model and QCA cells, wires, basic gates, and crossings are introduced. Four kinds of physical implementation for QCA are discussed with their advantages and disadvantages. Since most QCA device prototypes to date were demonstrated with semiconductor implementation, this book is based on the semiconductor QCA. However, the conclusions could also be applied to other implementation forms. Two types of clocking schemes, (i.e., quasi-adiabatic four-phase clocking and reversible Bennett clocking, are presented with their floorplans. Design and simulation tools that are extensively used in QCA research are discussed with a focus on the state-of-the-art simulation tool, QCADesigner. A survey of the QCA digital designs and testing methods proposed to date is presented. A set of basic design rules for QCA circuit design that should be followed in order to achieve robust designs is also discussed.

References

[1] Lent, C., et al., "Quantum Cellular Automata," *Nanotechnology*, Vol. 4, 1993, pp. 49–57.

[2] "QCA Home Page," website, 2013, http://www.nd.edu/~qcahome/.

[3] Walus, K., et al., "QCADesigner: A Rapid Design and Simulation Tool for Quantum-Dot Cellular Automata," *IEEE Transactions on Nanotechnology,* Vol. 3, 2004, pp. 26–31.

[4] Macucci, M., *Quantum Cellular Automata,* London: Imperial College Press, 2006.

[5] Lombardi, F., and J. Huang, *Design and Test of Digital Circuits by Quantum-Dot Cellular Automata,* Norwood, MA: Artech House, Inc., 2007.

[6] Tougaw, P., and C. Lent, "Logical Devices Implemented Using Quantum Cellular Automata," *Journal of Applied Physics,* Vol. 75, 1994, pp. 1818–1825.

[7] Frost, S., et al., "Memory in Motion: A Study of Storage Structures in QCA," in *Proceedings of 1st Workshop on Non-Silicon Computing,* Vol. 2, 2002, pp. 30–37.

[8] Walus, K., G. Schulhof, and G. Jullien, "High Level Exploration of Quantum-Dot Cellular Automata (QCA)," in *Conference Record of the 38th Asilomar Conference on Signals, Systems and Computers,* Vol. 1, 2004, pp. 30–33.

[9] Kim, K., K. Wu, and R. Karri, "Towards Designing Robust QCA Architectures in the Presence of Sneak Noise Paths," in *Proceedings of the Conference on Design, Automation and Test in Europe-Volume 2,* 2005, pp. 1214–1219.

[10] Kim, K., K. Wu, and R. Karri, "The Robust QCA Adder Designs Using Composable QCA Building Blocks," *IEEE Transactions on Computer-Aided Design of Integrated Circuits and Systems,* Vol. 26, 2007, pp. 176–183.

[11] Gin, A., P. Tougaw, and S. Williams, "An Alternative Geometry for Quantum-Dot Cellular Automata," *Journal of Applied Physics,* Vol. 85, 1999, pp. 8281–8286.

[12] Walus, K., and G. Jullien, "Design Tools for An Emerging SoC Technology: Quantum-Dot Cellular Automata," *Proceedings of the IEEE,* Vol. 94, 2006, pp. 1225–1244.

[13] Orlov, A., et al., "Realization of A Functional Cell for Quantum-Dot Cellular Automata," *Science,* Vol. 277, No. 5328, 1997, pp. 928–930.

[14] Bernstein, G., et al., "Observation of Switching in A Quantum-Dot Cellular Automata Cell," *Nanotechnology,* Vol. 10, 1999, pp. 166–173.

[15] Orlov, A., et al., "Correlated Electron Transport in Coupled Metal Double Dots," *Applied Physics Letters,* Vol. 73, 1998, pp. 2787–2789.

[16] Amlani, I., et al., "Digital Logic Gate Using Quantum-Dot Cellular Automata," *Science,* Vol. 284, No. 5412, 1999, pp. 289–291.

[17] Orlov, A., et al., "Experimental Demonstration of A Binary Wire for Quantum-Dot Cellular Automata," *Applied Physics Letters,* Vol. 74, No. 19, 1999, pp. 2875–2877.

[18] Amlani, I., et al., "Experimental Demonstration of A Leadless Quantum-Dot Cellular Automata Cell," *Applied Physics Letters,* Vol. 77, No. 5, 2000, pp. 738–740.

[19] Orlov, A., et al., "Experimental Demonstration of Clocked Single-Electron Switching in Quantum-Dot Cellular Automata," *Applied Physics Letters,* Vol. 77, No. 2, 2000, pp. 295–297.

[20] Khaetskii, A., and Y. Nazarov, "Spin Relaxation in Semiconductor Quantum Dots," *Physical Review B,* Vol. 61, No. 19, 2000, pp. 12639–12642.

[21] Single, C., et al., "Towards Quantum Cellular Automata Operation in Silicon: Transport Properties of Silicon Multiple Dot Structures," *Superlattices and Microstructures,* Vol. 28, No. 5, 2000, pp. 429–434.

[22] Smith, C., et al., "Realization of Quantum-Dot Cellular Automata Using Semiconductor Quantum Dots," *Superlattices and Microstructures,* Vol. 34, No. 3, 2003, pp. 195–203.

[23] Perez-Martinez, F., et al., "Demonstration of a Quantum Cellular Automata Cell in a GaAs/AlGaAs Heterostructure," *Applied Physics Letters,* Vol. 91, 2007, pp. 032 102(1–3).

[24] Walus, K., R. Budiman, and G. Jullien, "Impurity Charging in Semiconductor Quantum-Dot Cellular Automata," *Nanotechnology,* Vol. 16, 2005, pp. 2525–2529.

[25] Lent, C., "Bypassing the Transistor Paradigm," *Science,* Vol. 288, No. 5471, 2000, pp. 1597–1599.

[26] Lent, C., B. Isaksen, and M. Lieberman, "Molecular Quantum-Dot Cellular Automata," *Journal of the American Chemical Society,* Vol. 125, No. 4, 2003, pp. 1056–1063.

[27] Li, Z., A. Beatty, and T. Fehlner, "Molecular QCA Cells: 1. Structure and Functionalization of An Unsymmetrical Dinuclear Mixed-Valence Complex for Surface Binding," *Inorganic Chemistry,* Vol. 42, No. 18, 2003, pp. 5707–5714.

[28] Wang, Y., and M. Lieberman, "Thermodynamic Behavior of Molecular-Scale Quantum-Dot Cellular Automata (QCA) Wires and Logic Devices," *IEEE Transactions on Nanotechnology,* Vol. 3, 2004, pp. 368–376.

[29] Lu, Y., and C. Lent, "Theoretical Study of Molecular Quantum-Dot Cellular Automata," *Journal of Computational Electronics,* Vol. 4, No. 1, 2005, pp. 115–118.

[30] Haider, M., et al., "Controlled Coupling and Occupation of Silicon Atomic Quantum Dots at Room Temperature," *Physical Review Letters,* Vol. 102, No. 4, 2009, pp. 46 805–46 808.

[31] Hu, W., et al., "High-Resolution Electron Beam Lithography and DNA Nano-Patterning for Molecular QCA," *IEEE Transactions on Nanotechnology,* Vol. 4, 2005, pp. 312–316.

[32] Cowburn, R., and M. Welland, "Room Temperature Magnetic Quantum Cellular Automata," *Science,* Vol. 287, No. 5457, 2000, pp. 1466–1468.

[33] Bernstein, G., et al., "Magnetic QCA Systems," *Microelectronics Journal,* Vol. 36, No. 7, 2005, pp. 619–624.

[34] Imre, A., et al., "Majority Logic Gate for Magnetic Quantum-Dot Cellular Automata," *Science,* Vol. 311, No. 5758, 2006, pp. 205–208.

[35] Timler, J., and C. Lent, "Power Gain and Dissipation in Quantum-Dot Cellular Automata," *Journal of Applied Physics,* Vol. 91, 2002, pp. 823–831.

[36] Lent, C., and P. Tougaw, "A Device Architecture for Computing with Quantum Dots," *Proceedings of the IEEE,* Vol. 85, 1997, pp. 541–557.

[37] Frost, S., et al., "Carbon Nanotubes for Quantum-Dot Cellular Automata Clocking," in *Proceedings of the 4th IEEE Conference on Nanotechnology,* 2004, pp. 171–173.

[38] Alam, M., et al., "On-Chip Clocking for Nanomagnet Logic Devices," *IEEE Transactions on Nanotechnology,* Vol. 9, 2010, pp. 348–351.

[39] Frost-Murphy, S. E., et al., "On the Design of Reversible QDCA Systems," *Sandia National Laboratories Technical Report: SAND2006-5990*, 2006.

[40] Lent, C., P. Tougaw, and W. Porod, "Quantum Cellular Automata: the Physics of Computing with Arrays of Quantum Dot Molecules," in *Proceedings of Workshop on Physics and Computation*, 1994, pp. 5–13.

[41] Niemier, M., and P. Kogge, "Problems in Designing with QCAs: Layout= Timing," *International Journal of Circuit Theory and Applications*, Vol. 29, No. 1, 2001, pp. 49–62.

[42] Lent, C., M. Liu, and Y. Lu, "Bennett Clocking of Quantum-Dot Cellular Automata and the Limits to Binary Logic Scaling," *Nanotechnology*, Vol. 17, 2006, pp. 4240–4251.

[43] Bennett, C., "Logical Reversibility of Computation," *IBM Journal of Research and Development*, Vol. 17, 1973, pp. 525–532.

[44] Frost-Murphy, S., E. DeBenedictis, and P. Kogge, "General Floorplan for Reversible Quantum-Dot Cellular Automata," in *Proceedings of the 4th International Conference on Computing Frontiers*, 2007, pp. 77–82.

[45] Tougaw, P., and C. Lent, "Dynamic Behavior of Quantum Cellular Automata," *Journal of Applied Physics*, Vol. 80, 1996, pp. 4722–4736.

[46] Blair, E., "Tools for the Design and Simulation of Clocked Molecular Quantum-Dot Cellular Automata Circuits," Master's thesis, University of Notre Dame, Department of Electrical Engineering, 2003.

[47] Niemier, M., M. Kontz, and P. Kogge, "A Design of and Design Tools for A Novel Quantum Dot Based Microprocessor," in *Proceedings of the 37th Annual Design Automation Conference*, 2000, pp. 227–232.

[48] Tang, R., F. Zhang, and Y. Kim, "Quantum-Dot Cellular Automata SPICE Macro Model," in *Proceedings of the15th ACM Great Lakes Symposium on VLSI*, 2005, pp. 108–111.

[49] Ottavi, M., et al., "HDLQ: A HDL Environment for QCA Design," *ACM Journal on Emerging Technologies in Computing Systems*, Vol. 2, No. 4, 2006, pp. 243–261.

[50] Teodósio, T., and L. Sousa, "QCA-LG: A Tool for the Automatic Layout Generation of QCA Combinational Circuits," in *Proceedings of the Norchip*, 2007, pp. 1–5.

[51] Srivastava, S., et al., "QCAPro-An Error-Power Estimation Tool for QCA Circuit Design," in *Proceedings of the IEEE International Symposium on Circuits and Systems*, 2011, pp. 2377–2380.

[52] Walus, K., "QCADesigner," website, 2013, http://www.mina.ubc.ca/qcadesigner.

[53] Lent, C., and P. Tougaw, "Lines of Interacting Quantum-Dot Cells: A Binary Wire," *Journal of Applied Physics*, Vol. 74, No. 10, 1993, pp. 6227–6233.

[54] LaRue, M., D. Tougaw, and J. Will, "Stray charge in Quantum-Dot cellular automata: A validation of the intercellular hartree approximation," *IEEE Transactions on Nanotechnology*, Vol. 12, 2013, pp. 225–233.

[55] Taucer, M., et al., "Consequences of many-cell correlations in treating clocked Quantum-Dot cellular automata circuits," *arXiv preprint arXiv:1207.7008*, 2012.

[56] Timler, J., and C. Lent, "Maxwell's Demon and Quantum-Dot Cellular Automata," *Journal of Applied Physics*, Vol. 94, 2003, pp. 1050–1060.

[57] Bhanja, S., and S. Sarkar, "Probabilistic Modeling of QCA Circuits Using Bayesian Networks," *IEEE Transactions on Nanotechnology*, Vol. 5, 2006, pp. 657–670.

[58] Srivastava, S., and S. Bhanja, "Hierarchical Probabilistic Macromodeling for QCA Circuits," *IEEE Transactions on Computers*, Vol. 56, 2007, pp. 174–190.

[59] Srivastava, S., S. Sarkar, and S. Bhanja, "Estimation of Upper Bound of Power Dissipation in QCA Circuits," *IEEE Transactions on Nanotechnology*, Vol. 8, 2009, pp. 116–127.

[60] Walus, K., G. Jullien, and V. Dimitrov, "Computer Arithmetic Structures for Quantum Cellular Automata," in *Conference Record of the 37th Asilomar Conference on Signals, Systems and Computers*, Vol. 2, 2003, pp. 1435–1439.

[61] Hänninen, I., and J. Takala, "Arithmetic Design on Quantum-Dot Cellular Automata Nanotechnology," in *Proceedings of the 8th International Workshop on Embedded Computer Systems: Architectures, Modeling, and Simulation*, 2008, pp. 43–52.

[62] Swartzlander, Jr., E., et al., "Computer Arithmetic Implemented with QCA: A Progress Report," in *Conference Record of the 44th Asilomar Conference on Signals, Systems and Computers*, 2010, pp. 1392–1398.

[63] Wang, W., K. Walus, and G. Jullien, "Quantum-Dot Cellular Automata Adders," in *Proceedings of the 3rd IEEE Conference on Nanotechnology*, Vol. 1, 2003, pp. 461–464.

[64] Zhang, R., et al., "Performance Comparison of Quantum-Dot Cellular Automata Adders," in *Proceedings of the IEEE International Symposium on the Circuits and Systems*, 2005, pp. 2522–2526.

[65] Hänninen, I., and J. Takala, "Robust Adders Based Quantum-Dot Cellular Automata," in *Proceedings of the IEEE International Conference on Application-specific Systems, Architectures and Processors*, 2007, pp. 391–396.

[66] Hänninen, I., and J. Takala, "Binary Adders on Quantum-Dot Cellular Automata," *Journal of Signal Processing Systems,*, Vol. 58, No. 1, 2010, pp. 87–103.

[67] Cho, H., and E. Swartzlander, Jr., "Adder and Multiplier Design in Quantum-Dot Cellular Automata," *IEEE Transactions on Computers*, Vol. 58, 2009, pp. 721–727.

[68] Cho, H., and E. Swartzlander, Jr., "Pipelined Carry Lookahead Adder Design in Quantum-Dot Cellular Automata," in *Conference Record of the 39th Asilomar Conference on Signals, Systems and Computers*, 2005, pp. 1191–1195.

[69] Cho, H., and E. Swartzlander, Jr., "Modular Design of Conditional Sum Adders Using Quantum-Dot Cellular Automata," in *Proceedings of the 6th IEEE Conference on Nanotechnology*, Vol. 1, 2006, pp. 363–366.

[70] Cho, H., and E. Swartzlander, Jr., "Adder Designs and Analyses for Quantum-Dot Cellular Automata," *IEEE Transactions on Nanotechnology*, Vol. 6, 2007, pp. 374–383.

[71] Pudi, V., and K. Sridharan, "Efficient Design of A Hybrid Adder in Quantum-Dot Cellular Automata," *IEEE Transactions on Very Large Scale Integration Systems*, Vol. 19, 2011, pp. 1535– 1548.

[72] Pudi, V., and K. Sridharan, "Low Complexity Design of Ripple Carry and Brent-Kung Adders in QCA," *IEEE Transactions on Nanotechnology*, Vol. 11, 2012, pp. 105–119.

[73] Hänninen, I., and J. Takala, "Binary Multipliers on Quantum-Dot Cellular Automata," *Facta Universitatis-Series: Electronics and Energetics,* Vol. 20, No. 3, 2007, pp. 541–560.

[74] Cho, H., and E. Swartzlander, Jr., "Serial Parallel Multiplier Design in Quantum-Dot Cellular Automata," in *Proceedings of the 18th IEEE Symposium on Computer Arithmetic,* 2007, pp. 7–15.

[75] Kim, S., and E. Swartzlander, Jr., "Parallel Multipliers for Quantum-Dot Cellular Automata," in *Proceedings of the IEEE Nanotechnology Materials and Devices Conference,* 2009, pp. 68–72.

[76] Hänninen I., J. Takala, "Pipelined Array Multiplier Based Quantum-Dot Cellular Automata," in *Proceedings of the 18th European Conference on Circuit Theory and Design,* 2007, pp. 938–941.

[77] Kim, S., and E. Swartzlander, Jr., "Multipliers with Coplanar Crossings for Quantum-Dot Cellular Automata," in *Proceedings of 10th IEEE Conference on Nanotechnology,* 2010, pp. 953–957.

[78] Hänninen, I., and J. Takala, "Radix-4 Recoded Multiplier on Quantum-Dot Cellular Automata," in *Proceedings of the 9th International Workshop on Embedded Computer Systems: Architectures, Modeling, and Simulation,* 2009, pp. 118–127.

[79] Kong, I., E. Swartzlander, Jr., and S. Kim, "Design of a Goldschmidt Iterative Divider for Quantum-Dot Cellular Automata," in *Proceedings of the IEEE/ACM International Symposium on Nanoscale Architectures,* 2009, pp. 47–50.

[80] Kim, S., and E. Swartzlander Jr., "Restoring Divider Design for Quantum-Dot Cellular Automata," in *Proceedings of the 11th IEEE Conference on Nanotechnology,* 2011, pp. 1295–1300.

[81] Wood, J., and D. Tougaw, "Matrix Multiplication Using Quantum-Dot Cellular Automata to Implement Conventional Microelectronics," *IEEE Transactions on Nanotechnology,* Vol. 10, 2011, pp. 1036–1042.

[82] Lu, L., et al., "QCA Systolic Matrix Multiplier," in *Proceedings of the IEEE Computer Society Annual Symposium on VLSI,* 2010, pp. 149–154.

[83] Lu, L., et al., "QCA Systolic Array Design," *IEEE Transactions on Computers,* Vol. 62, 2013, pp. 548–560.

[84] Liu, W., et al., "Design Rules for Quantum-Dot Cellular Automata," in *Proceedings of the IEEE International Symposium on Circuits and Systems,* 2011, pp. 2361–2364.

[85] Liu, W., et al., "Montgomery Modular Multiplier Design in Quantum-Dot Cellular Automata using Cut-Set Retiming," in *Proceedings of the 10th IEEE Conference on Nanotechnology,* 2010, pp. 205–210.

[86] Liu, W., et al., "Design of Quantum-Dot Cellular Automata Circuits Using Cut-Set Retiming," *IEEE Transactions on Nanotechnology,* Vol. 10, 2011, pp. 1150–1160.

[87] Taghizadeh, M., M. Askari, and K. Fardad, "BCD Computing Structures in Quantum-Dot Cellular Automata," in *Proceedings of the IEEE International Conference on Computer and Communication Engineering,* 2008, pp. 1042–1045.

[88] Kharbash, F., and G. Chaudhry, "The Design of Quantum-Dot Cellular Automata Decimal Adder," in *Proceedings of the IEEE International Multitopic Conference,* 2008, pp. 71–75.

[89] Gladshtein, M., "Quantum-Dot Cellular Automata Serial Decimal Adder," *IEEE Transactions on Nanotechnology,* Vol. 10, 2011, pp. 1377–1382.

[90] Liu, W., et al., "Cost-Efficient Decimal Adder Design in Quantum-Dot Cellular Automata," in *Proceedings of the IEEE International Symposium on Circuits and Systems,* 2012, pp. 1347–1350.

[91] Momenzadeh, M., et al., "Characterization, Test, and Logic Synthesis of And-Or-Inverter (AOI) Gate Design for QCA Implementation," *IEEE Transactions on Computer-Aided Design of Integrated Circuits and Systems,* Vol. 24, 2005, pp. 1881–1893.

[92] Gin, A., et al., "Hierarchical Design of Quantum-Dot Cellular Automata Devices," *Journal of Applied Physics,* Vol. 85, 1999, pp. 3713–3720.

[93] Teja, V., S. Polisetti, and S. Kasavajjala, "QCA Based Multiplexing of 16 Arithmetic & Logical Subsystems-A Paradigm for Nano Computing," in *Proceedings of the 3rd IEEE International Conference on Nano/Micro Engineered and Molecular Systems,* 2008, pp. 758–763.

[94] Mardiris, V., and I. Karafyllidis, "Design and Simulation of Modular 2n to 1 Quantum-Dot Cellular Automata (QCA) Multiplexers," *International Journal of Circuit Theory and Applications,* Vol. 38, No. 8, 2010, pp. 771–785.

[95] Vetteth, A., et al., "Quantum-Dot Cellular Automata Carry-Lookahead Adder and Barrel Shifter," in *Proceedings of the IEEE Emerging Telecommunications Technologies Conference,* 2002, pp. 1–5.

[96] Huang, J., et al., "Tile-Based QCA Design Using Majority-Like Logic Primitives," *ACM Journal on Emerging Technologies in Computing Systems,* Vol. 1, No. 3, 2005, pp. 163–185.

[97] Vankamamidi, V., M. Ottavi, and F. Lombardi, "Two-Dimensional Schemes for Clocking/Timing of QCA Circuits," *IEEE Transactions on Computer-Aided Design of Integrated Circuits and Systems,* Vol. 27, No. 1, 2008, pp. 34–44.

[98] Huang, J., M. Momenzadeh, and F. Lombardi, "Design of Sequential Circuits by Quantum-Dot Cellular Automata," *Microelectronics Journal,* Vol. 38, No. 4-5, 2007, pp. 525–537.

[99] Askari, M., M. Taghizadeh, and K. Fardad, "Design and Analysis of A Sequential Ring Counter for QCA Implementation," in *Proceedings of the International Conference on Computer and CommunicationEngineering,* 2008, pp. 933–936.

[100] Shamsabadi, A., et al., "Applying Inherent Capabilities of Quantum-Dot Cellular Automata to Design: D Flip-Flop Case Study," *Journal of Systems Architecture,* Vol. 55, No. 3, 2009, pp. 180–187.

[101] Yang, X., L. Cai, and X. Zhao, "Low Power Dual-Edge Triggered Flip-Flop Structure in Quantum Dot Cellular Automata," *Electronics Letters,* Vol. 46, No. 12, 2010, pp. 825–826.

[102] Venkataramani, P., S. Srivastava, and S. Bhanja, "Sequential Circuit Design in Quantum-Dot Cellular Automata," in *Proceedings of the 8th IEEE Conference on Nanotechnology*, 2008, pp. 534–537.

[103] Yang, X., et al., "Design and Simulation of Sequential Circuits in Quantum-Dot Cellular Automata: Falling Edge-Triggered Flip-Flop and Counter Study," *Microelectronics Journal*, Vol. 41, No. 1, 2010, pp. 56–63.

[104] Thapliyal, H., and N. Ranganathan, "Reversible Logic-Based Concurrently Testable Latches for Molecular QCA," *IEEE Transactions on Nanotechnology*, Vol. 9, 2010, pp. 62–69.

[105] Berzon, D., and T. Fountain, "A Memory Design in QCAs Using the SQUARES Formalism," in *Proceedings of 9th Great Lakes Symposium on VLSI*, 1999, pp. 166–169.

[106] Walus, K., et al., "RAM Design Using Quantum-Dot Cellular Automata," in *Proceedings of Nanotechnology Conference*, Vol. 2, 2003, pp. 160–163.

[107] Vankamamidi, V., M. Ottavi, and F. Lombardi, "A Serial Memory by Quantum-Dot Cellular Automata (QCA)," *IEEE Transactions on Computers*, Vol. 57, 2008, pp. 606–618.

[108] Dehkordi, M., et al., "Novel RAM Cell Designs Based on Inherent Capabilities of Quantum-Dot Cellular Automata," *Microelectronics Journal*, Vol. 42, 2011, pp. 701–708.

[109] Vankamamidi, V., M. Ottavi, and F. Lombardi, "A Line-Based Parallel Memory for QCA Implementation," *IEEE Transactions on Nanotechnology*, Vol. 4, 2005, pp. 690–698.

[110] Taskin, B., and B. Hong, "Improving Line-Based QCA Memory Cell Design Through Dual Phase Clocking," *IEEE Transactions on Very Large Scale Integration Systems*, Vol. 16, 2008, pp. 1648–1656.

[111] Taskin, B., et al., "A Shift-Register-Based QCA Memory Architecture," *ACM Journal on Emerging Technologies in Computing Systems*, Vol. 5, No. 1, 2009, pp. 4:1–18.

[112] Niemier, M., and P. Kogge, "Logic in Wire: Using Quantum Dots to Implement A Microprocessor," in *Proceedings of the 6th IEEE International Conference on Electronics, Circuits and Systems*, Vol. 3, 1999, pp. 1211–1215.

[113] Walus, K., et al., "Simple 4-bit Processor Based on Quantum-Dot Cellular Automata (QCA)," in *Proceedings of the 16th IEEE International Conference on Application-Specific Systems, Architecture Processors*, 2005, pp. 288–293.

[114] Niemier, M., A. Rodrigues, and P. Kogge, "A Potentially Implementable FPGA for Quantum-Dot Cellular Automata," in *Proceedings of the 1st Workshop on Non-silicon Computation*, 2002, pp. 38–45.

[115] Crocker, M., et al., "PLAs in Quantum-Dot Cellular Automata," *IEEE Transactions on Nanotechnology*, Vol. 7, 2008, pp. 376–386.

[116] Amiri, M., M. Mahdavi, and S. Mirzakuchaki, "QCA Implementation of A MUX-Based FPGA CLB," in *Proceedings of the International Conference on Nanoscience and Nanotechnology*, 2008, pp. 141–144.

[117] Tung, C., R. Rungta, and E. Peskin, "Simulation of A QCA-Based CLB and A Multi-CLB Application," in *Proceedings of the International Conference on Field-Programmable Technology*, 2009, pp. 62–69.

[118] Helsingius, M., P. Kuosmanen, and J. Astola, "Nonlinear Filters Using Quantum-Dot Cells," *Electronics Letters*, Vol. 33, No. 20, 1997, pp. 1735–1736.

[119] Fountain, T., "The Design of Highly-Parallel Image Processing Systems Using Nanoelectronic Devices," in *Proceedings of the 4th IEEE International Workshop on Computer Architecture for Machine Perception*, 1997, pp. 210–219.

[120] Cardenas-Barrera, J., K. N. Plataniotis, and A. Venetsanopoulos, "QCA Implementation of A Multichannel Filter for Image Processing," *Mathematical Problems in Engineering*, Vol. 8, No. 1, 2002, pp. 87–99.

[121] Amiri, M., M. Mahdavi, and S. Mirzakuchaki, "QCA Implementation of A5/1 Stream Cipher," in *Proceedings of the 2nd International Conference on Advances in Circuits, Electronics and Micro-electronics*, 2009, pp. 48–51.

[122] Amiri, M., et al., "QCA Implementation of Serpent Block Cipher," in *Proceedings of the 2nd International Conference on Advances in Circuits, Electronics and Micro-electronics*, 2009, pp. 16–19.

[123] Choi, M., et al., "Designing Layout-Timing Independent Quantum-Dot Cellular Automata (QCA) Circuits by Global Asynchrony," *Journal of Systems Architecture*, Vol. 53, No. 9, 2007, pp. 551– 567.

[124] Graziano, M., et al., "A NCL-HDL Snake-Clock Based Magnetic QCA Architecture," *IEEE Transactions on Nanotechnology*, Vol. 10, 2011, pp. 1141–1149.

[125] Graziano, M., et al., "Asynchrony in Quantum-Dot Cellular Automata Nanocomputation: Elixir or Poison?" *IEEE Design and Test of Computers*, Vol. 28, 2011, pp. 72–83.

[126] Zhang, R., et al., "A Method of Majority Logic Reduction for Quantum Cellular Automata," *IEEE Transactions on Nanotechnology*, Vol. 3, 2004, pp. 443–450.

[127] Huo, Z., et al., "Logic Optimization for Majority Gate-Based Nanoelectronic Circuits," in *Proceedings of the IEEE International Symposium on Circuits and Systems*, 2006, pp. 1307–1310.

[128] Bonyadi, M., et al., "Logic Optimization for Majority Gate-Based Nanoelectronic Circuits Based on Genetic Algorithm," in *Proceedings of the International Conference on Electrical Engineering*, 2007, pp. 1–5.

[129] Kong, K., Y. Shang, and R. Lu, "An Optimized Majority Logic Synthesis Methodology for Quantum-Dot Cellular Automata," *IEEE Transactions on Nanotechnology*, Vol. 9, 2010, pp. 170–183.

[130] Zhang, R., P. Gupta, and N. Jha, "Synthesis of Majority and Minority Networks and Its Applications to QCA-, TPL-and SET-Based Nanotechnologies," in *Proceedings of the 18th International Conference on VLSI Design*, 2005, pp. 229–234.

[131] Zhang, R., P. Gupta, and N. Jha, "Majority and Minority Network Synthesis with Application to QCA-, SET-, and TPL-Based Nanotechnologies," *IEEE Transactions on Computer-Aided Design of Integrated Circuits and Systems*, Vol. 26, 2007, pp. 1233–1245.

[132] Gergel, N., S. Craft, and J. Lach, "Modeling QCA for Area Minimization in Logic Synthesis," in *Proceedings of the 13th ACM Great Lakes Symposium on VLSI*, 2003, pp. 60–63.

[133] Lim, S., R. Ravichandran, and M. Niemier, "Partitioning and Placement for Buildable QCA Circuits," *ACM Journal on Emerging Technologies in Computing Systems,* Vol. 1, No. 1, 2005, pp. 50–72.

[134] Bubna, M., et al., "A Layout-Aware Physical Design Method for Constructing Feasible QCA Circuits," in *Proceedings of the 18th ACM Great Lakes Symposium on VLSI,* 2008, pp. 243–248.

[135] Dysart, T., and P. Kogge, "Strategy and Prototype Tool for Doing Fault Modeling in A Nanotechnology," in *Proceedings of the 3rd IEEE Conference on Nanotechnology,* Vol. 1, 2003, pp. 356–359.

[136] Momenzadeh, M., M. Ottavi, and F. Lombardi, "Modeling QCA Defects at Molecular-Level in Combinational Circuits," in *Proceedings of the 20th IEEE International Symposium on Defect and Fault Tolerancein VLSI Systems,* 2005, pp. 208–216.

[137] Tahoori, M., et al., "Testing of Quantum Cellular Automata," *IEEE Transactions on Nanotechnology,* Vol. 3, 2004, pp. 432–442.

[138] Momenzadeh, M., et al., "Quantum Cellular Automata: New Defects and Faults for New Devices," in *Proceedings of the 18th International Symposium on Parallel and Distributed Processing Symposium,* 2004, pp. 207–214.

[139] Khatun, M., et al., "Fault Tolerance Properties in Quantum-Dot Cellular Automata Devices," *Journal of Physics D: Applied Physics,* Vol. 39, 2006, pp. 1489–1494.

[140] Momenzadeh, M., et al., "On the Evaluation of Scaling of QCA Devices in the Presence of Defects at Manufacturing," *IEEE Transactions on Nanotechnology,* Vol. 4, 2005, pp. 740–743.

[141] Huang, J., M. Momenzadeh, and F. Lombardi, "Defect Tolerance of QCA Tiles," in *Proceedings of the Design, Automation and Test in Europe Conference,* Vol. 1, 2006, pp. 1–6.

[142] Hänninen, I., and J. Takala, "Reliability of N-Bit Nanotechnology Adder," in *Proceedings of the IEEE Computer Society Annual Symposium on VLSI,* 2008, pp. 34–39.

[143] Hänninen, I., and J. Takala, "Reliability of A QCA Array Multiplier," in *Proceedings of the 8th IEEE Conference on Nanotechnology,* 2008, pp. 315–318.

[144] Momenzadeh, M., J. Huang, and F. Lombardi, "Defect Characterization and Tolerance of QCA Sequential Devices and Circuits," in *Proceedings of the 20th IEEE International Symposium on Defect and Fault Tolerance in VLSI Systems,* 2005, pp. 199–207.

[145] Huang, J., M. Momenzadeh, and F. Lombardi, "Analysis of Missing and Additional Cell Defects in Sequential Quantum-Dot Cellular Automata," *Integration, the VLSI Journal,* Vol. 40, No. 4, 2007, pp. 503–515.

[146] Karim, F., and K. Walus, "Characterization of the Displacement Tolerance of QCA Interconnects," in *Proceedings of the IEEE International Workshop on Design and Test of NanoDevices, Circuits and Systems,* 2008, pp. 49–52.

[147] Crocker, M., X. Hu, and M. Niemier, "Defects and Faults in QCA-Based PLAs," *ACM Journal on Emerging Technologies in Computing Systems,* Vol. 5, No. 2, 2009, pp. 8:1–27.

[148] Dai, J., L. Wang, and F. Lombardi, "An Information-Theoretic Analysis of Quantum-Dot Cellular Automata for Defect Tolerance," *ACM Journal on Emerging Technologies in Computing Systems,* Vol. 6, No. 3, 2010, pp. 9:1–19.

[149] Ma, X., et al., "Testing Reversible 1D Arrays for Molecular QCA," in *Proceedings of the 21st IEEE International Symposium on Defect and Fault Tolerance in VLSI Systems,* 2006, pp. 71–79.

[150] Ma, X., et al., "Reversible Gates and Testability of One Dimensional Arrays of Molecular QCA," *Journal of Electronic Testing,* Vol. 24, No. 1, 2008, pp. 297–311.

[151] Ma, X., et al., "Reversible and Testable Circuits for Molecular QCA Design," *Emerging Nanotechnologies,* 2008, pp. 157–202.

[152] Niemier, M., R. Ravichandran, and P. Kogge, "Using Circuits and Systems-Level Research to Drive Nanotechnology," in *Proceedings of the IEEE International Conference on Computer Design: VLSI in Computers and Processors,* 2004, pp. 302–309.

[153] Shukla, S., and R. Bahar, *Nano, Quantum and Molecular Computing: Implications to High Level Design and Validation,* Norwell, MA: Kluwer Academic Publishers, 2004.

[154] Bernstein, G., "Quantum-Dot Cellular Automata: Computing by Field Polarization," in *Proceedings of the 40th Annual Design Automation Conference,* 2003, pp. 268–273.

[155] Tóth, G., and C. S. Lent, "Role of correlation in the operation of Quantum-Dot cellular automata," *Journal of Applied Physics,* Vol. 89, No. 12, 2001, pp. 7943–7953.

Part II
QCA Arithmetic Circuits

3

QCA Adders

Heumpil Cho, Earl E. Swartzlander, Jr., Weiqiang Liu, and Máire O'Neill

3.1 Introduction

This chapter investigates the design and implementation of binary and decimal adders. Adders are important because they form the basis for most arithmetic units. The chapter is organized in five sections. The first three sections show the design in QCA of conventional CMOS adders (RCAs, CLAs and CSAs). Not surprisingly, the QCA design of the CLA offers better performance than the QCA RCA. Section 3.5 summarizes the characteristics of the conventional adders. Section 3.6 shows the design of a modification of the traditional RCA called the CFA. By optimizing the design of a full adder to take advantage of QCA characteristics, the CFA achieves better performance than the CLA, even for large word sizes. Finally, Section 3.7 shows the design of decimal adders in QCA.

3.2 Ripple Carry Adder[1]

3.2.1 Architectural Design

The first adder is the RCA. In conventional integrated circuits this adder is simple, small, and slow.

There are already proposed designs for RCAs in QCA [2, 3]. In this section, the design is extended to pipeline structures for comparison with CLAs and CSAs. For input and output synchronization, a series of wires with extra clock zones is added. The design shows a staircase-like design due to the extra

1. Section 3.2 is based on [1].

wire channels. The design can be adjusted to accommodate any operand size by appending extra full adders.

3.2.2 Schematic Design

The following equations are used for a full adder:

$$s_i = a_i b_i c_i + a_i \overline{b_i} \overline{c_i} + \overline{a_i} b_i \overline{c_i} + \overline{a_i} \overline{b_i} c_i \tag{3.1}$$

$$s_i = M\left(\overline{M}\left(a_i, b_i, c_i\right), M\left(a_i, b_i, \overline{c_i}\right), c_i\right) \tag{3.2}$$

$$c_{i+1} = a_i b_i + b_i c_i + a_i c_i \tag{3.3}$$

$$c_{i+1} = M\left(a_i, b_i, c_i\right) \tag{3.4}$$

where, a_i and b_i denote inputs at bit position i. s_i represents the sum output at bit position i and c_{i+1} represents the carry output generated from bit position i. Using these equations, the gate level design is implemented as shown in Figure 3.1. RCAs are made by series connections of full adders. The carry input for each full adder is the carry output from the adjacent lower full adder.

3.2.3 Layout Design

Figure 3.2 shows a full adder layout using multilayer crossovers. The full adder takes one clock cycle to form the sum and carry outputs. The carries ripple from stage to stage from the least significant bit to the most significant bit.

Extra wire channels are added to the RCA input and output sides for synchronization. An n-bit adder has a delay of n clock cycles. A 4-bit RCA layout is shown in Figure 3.3.

3.3 Carry Lookahead Adder[2]

3.3.1 Architectural Design

The CLA has a regular structure. It achieves high speed with a moderate complexity. For this section, 4-, 8-, 16-, 32-, and 64-bit CLA were developed following basic CMOS pipelined adder designs [5]. The pipelined designs avoid feedback signals (used in regular CMOS CLAs) that are difficult to implement

---

2. Section 3.3 is based on [1, 4].

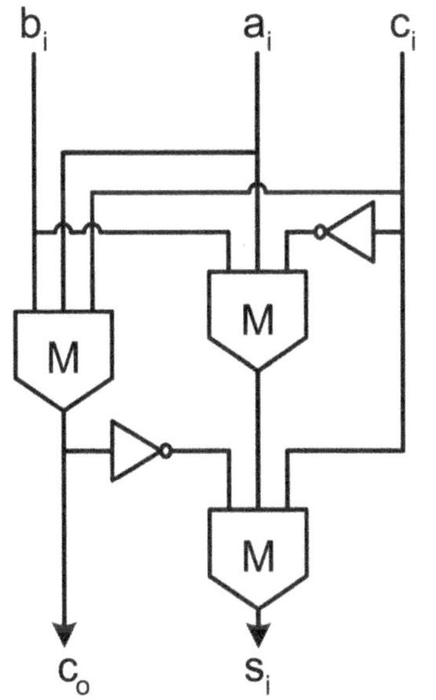

Figure 3.1 1-bit full adder schematic. (© 2007 IEEE. *From* [1].)

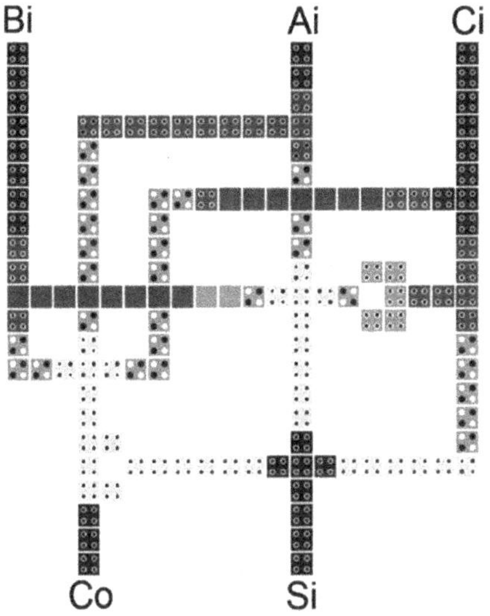

Figure 3.2 1-bit full adder layout. (© 2007 IEEE. *From* [1].)

with QCAs. Figures 3.4–3.6 show block diagrams of the designs for 4, 16, and 64-bit CLAs, respectively. The designs employ 4-bit slices for the lookahead logic, so each increase in word size by a factor of 4 requires an additional level of lookahead logic.

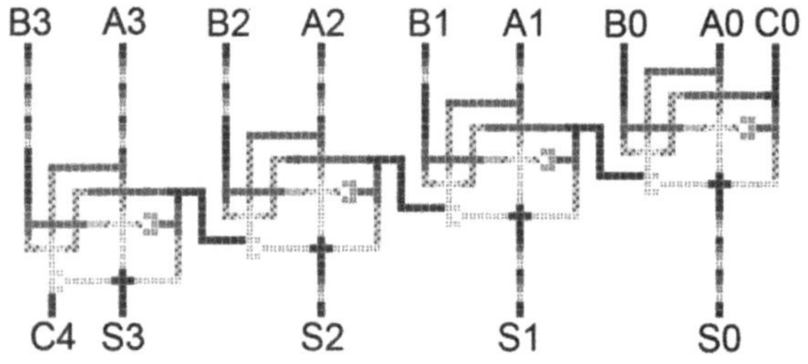

Figure 3.3 4-bit RCA layout. (© 2007 IEEE. *From* [1].)

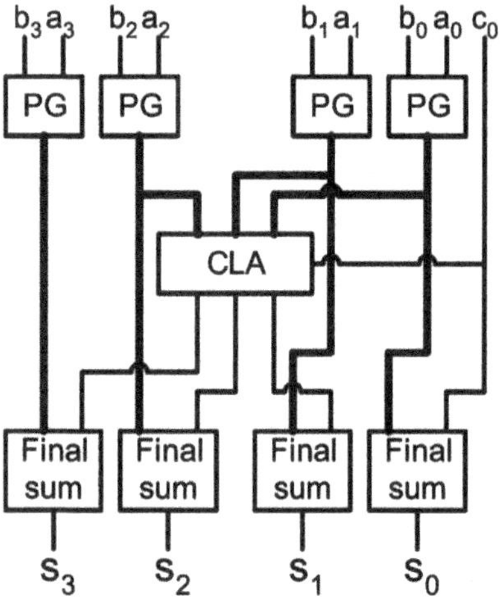

Figure 3.4 4-bit CLA block diagram. (© 2007 IEEE. *From* [1].)

The PG block has a generate output, $g_i = a_i b_i$ that indicates that a carry is "generated" at bit position i and a propagate output $p_i = a_i + b_i$ that indicates that a carry entering bit position i will "propagate" to the next bit position. They are used to produce all the carries in parallel at the successive blocks. The block PG section produces and transfers block generate/propagate signals to the next higher level.

The CLA and block CLA sections are virtually identical except for the different hierarchy of their positions and additional bypassing signals. Their outputs and PG outputs are used to calculate the final sum at each bit position. Due to the pipeline design, all sum output signals are available at the same clock period.

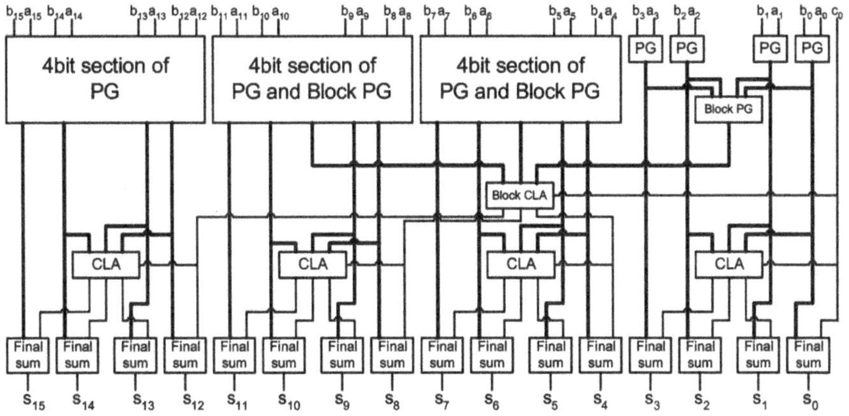

Figure 3.5 16-bit CLA block diagram. (© 2007 IEEE. *From* [1].)

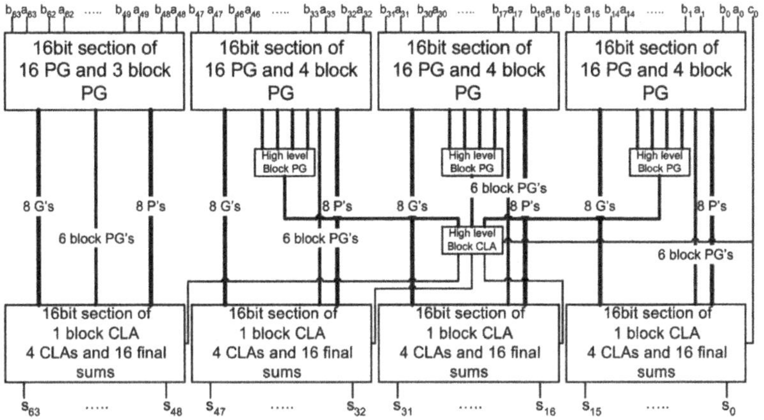

Figure 3.6 64-bit CLA block diagram. (© 2007 IEEE. *From* [1].)

3.3.2 Schematic Design

Using the block PG equations, AND/OR logic functions are mapped to majority gates to build the 4-bit section of PG and block PG. The CLA and block CLA sections are described by the following equations:

$$p_b = p_{i+3} + p_{i+2} + p_{i+1} + 1p_i \tag{3.5}$$

$$g_b = g_{i+3} + p_{i+3}g_{i+2} + p_{i+3}p_{i+2}g_{i+1} + p_{i+3}p_{i+2}p_{i+1}g_i \tag{3.6}$$

$$c_{i+1} = g_i + p_i c_i \tag{3.7}$$

$$c_{i+2} = g_{i+1} + p_{i+1} g_i + p_{i+1} p_i c_i \tag{3.8}$$

$$c_{i+3} = g_{i+2} + p_{i+2} g_{i+1} + p_{i+2} p_{i+1} g_i + p_{i+2} p_{i+1} p_i c_i \tag{3.9}$$

Figures 3.7 and 3.8 show the corresponding gate level block diagrams. Due to the characteristics of majority gates in QCA, a half adder and a full adder have the same complexity. The full adder design [3] uses three majority gates. An Exclusive OR gate (equivalent to a half adder) also needs three majority gates. Thus, both adders have the same complexity ignoring the wire routing complexity. The final sum adder is similar to a full adder except that the inputs are p_i, g_i, and c_i rather than a_i, b_i, and c_i. Using those inputs, wire routing is simplified. Table 3.1 shows the Karnaugh map of those three inputs for Boolean

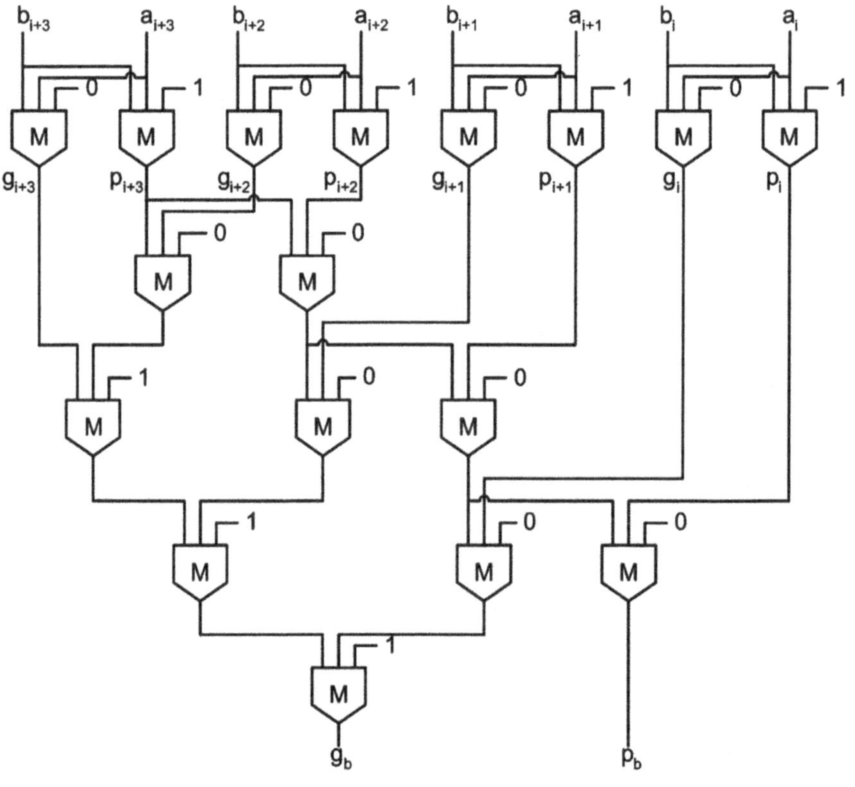

Figure 3.7 4-bit section of PG and block PG schematic. (© 2007 IEEE. *From* [1].)

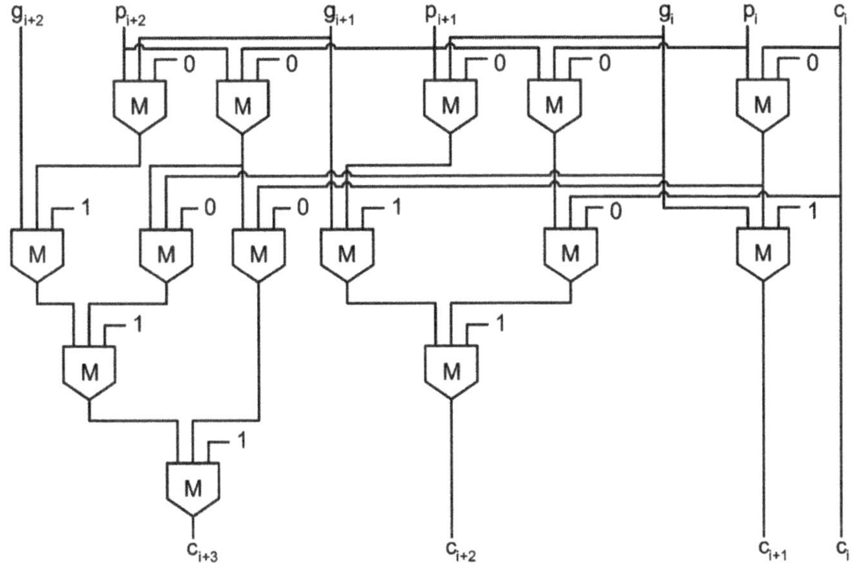

Figure 3.8 4-bit section of CLA and block CLA schematic. (© 2007 IEEE. *From* [1].)

Table 3.1
Karnaugh Map of Final Sum Adder

$p_i g_i$	00	01	11	10
c_i				
0	0	X(1)	0	1
1	1	X(0)	1	0

optimization. A generate signal is always included in a propagate signal. There are two don't care sets ($p_i = 0$, $g_i = 1$) in the Karnaugh map. The logic values, in parenthesis, were used for the optimization.

After the optimization, the final sum adder is implemented with three majority gates. Figure 3.9 shows the gate level diagram of the given final sum adder equation.

$$s_i = p_i g_i c_i + p_i \overline{g}_i \overline{c}_i + \overline{p}_i g_i \overline{c}_i + \overline{p}_i \overline{g}_i c_i \tag{3.10}$$

$$s_i = M\left(M\left(\overline{p}_i, g_i, c_i\right), M\left(p_i, \overline{g}_i, c_i\right) \overline{c}_i\right) \tag{3.11}$$

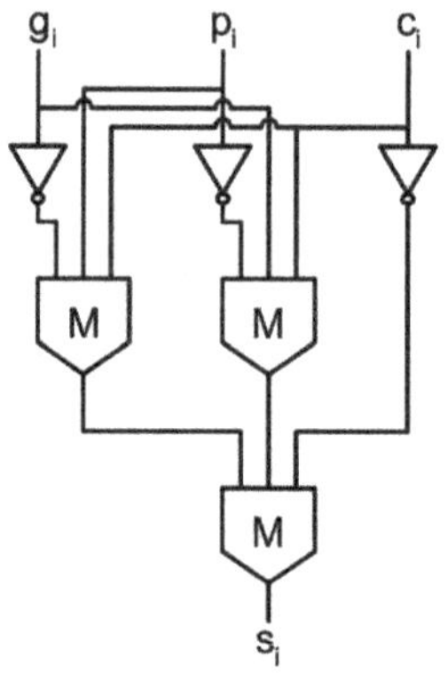

Figure 3.9 Final sum adder schematic. (© 2007 IEEE. *From* [1].)

3.3.3 Layout Design

Figures 3.10 and 3.11 show the layouts of 4 and 16-bit CLAs from QCADesigner.

3.3.4 Simulation Results

With QCADesigner [6], the circuit functionality of the CLA is verified. In the simulations, most default parameters for a bistable approximation as listed in Section 2.4.1.3 are used. The number of samples is determined to be 102,400.

For clarity, only the 4-bit CLA simulation results are shown. The input and output waveforms are shown in Figure 3.12. The first meaningful output appears in the fourth clock tick after 3.5 clock delays. The first and last input/output pairs are highlighted.

3.4 Conditional Sum Adder[3]

3.4.1 Architectural Design

In CMOS, the conditional sum adder (CSA) is frequently used when the highest speed is required. CSAs of 4, 8, 16, 32, and 64 bits were designed and simulated. The structures are based on the recursive relations shown in Figure 3.13.

This design can be divided into two half-size calculations. The upper half calculation is duplicated (one assuming a carry-in of 0 and one assuming a carry-in of 1). The carry output from the lower half is used to select the correct upper half output. This process is continued recursively down to the bit level.

These recursive relations produce modular designs. Figure 3.14 shows the block diagram of an 8-bit CSA. The blocks just below the modified half adders

3. Section 3.4 is based on [1, 7].

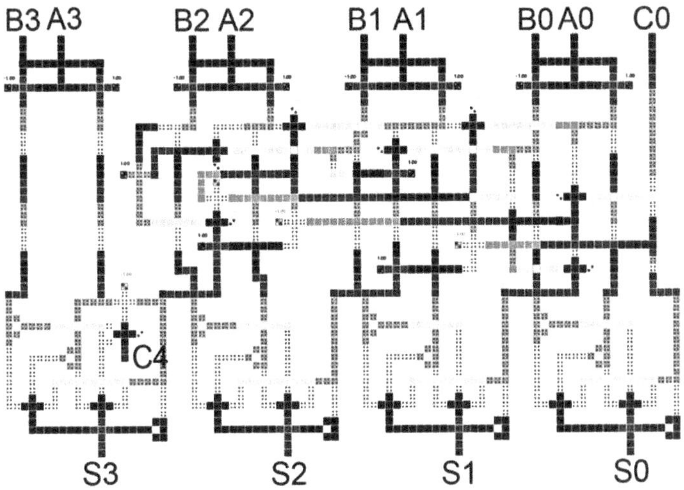

Figure 3.10 4-bit CLA layout. (© 2007 IEEE. *From* [1].)

(MHAs) are referred to as level one. Successive lower blocks are called level two, three, and so on.

3.4.2 Schematic Design

The following equations are used for CSAs. a_i and b_i denote inputs at bit position i. s_i represents the sum output at bit position i and c_{i+1} represents the carry output generated from bit position i. s_i^p means the sum of bit position i when the carry input value is p, and c_{i+1}^q means the carry output of bit position i when the carry input is q. At bit position i, the definitions of sum and carry output are:

$$s_i^0 = a_i \oplus b_i, s_i^1 = a_i \overline{\oplus} b_i = \overline{a_i \oplus b_i} = \overline{s_i^0}, c_{i+1}^0 = a_i b_i, \text{and } c_{i+1}^1 = a_i + b_i$$

$$s_0 = a_0 \oplus b_0 \oplus c_0 \tag{3.12}$$

$$c_1 = M\left(a_0, b_0, c_0\right) \tag{3.13}$$

$$s_1 = s_1^0 \overline{c_1} + s_1^1 c_1 \tag{3.14}$$

$$c_2 = c_2^0 \overline{c_1} + c_2^1 c_1 \tag{3.15}$$

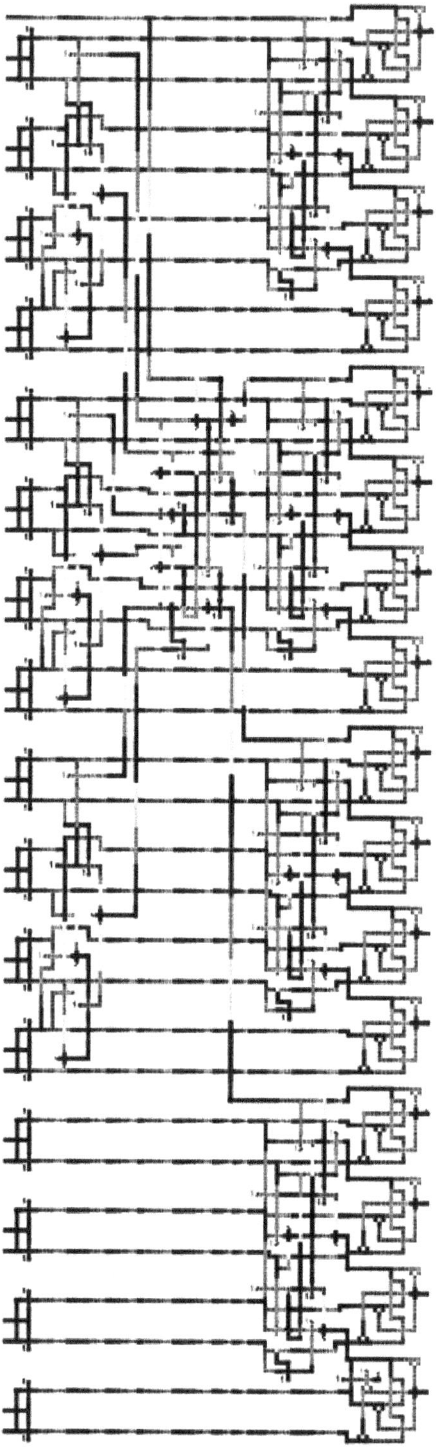

Figure 3.11 16-bit CLA layout. (© 2007 IEEE. *From* [1].)

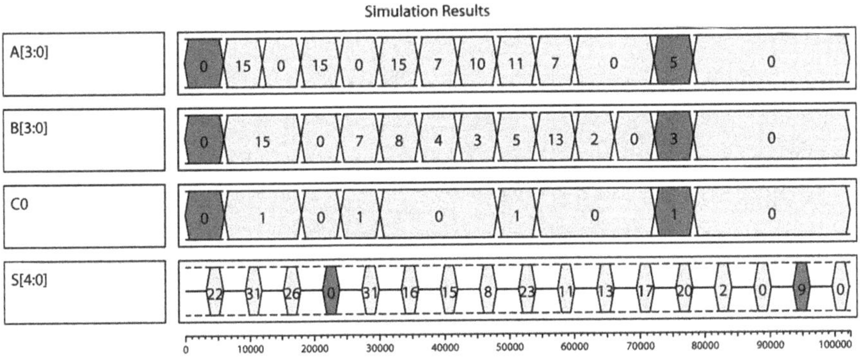

Figure 3.12 Simulation results for 4-bit CLA using QCADesigner. (© 2007 IEEE. *From* [1].)

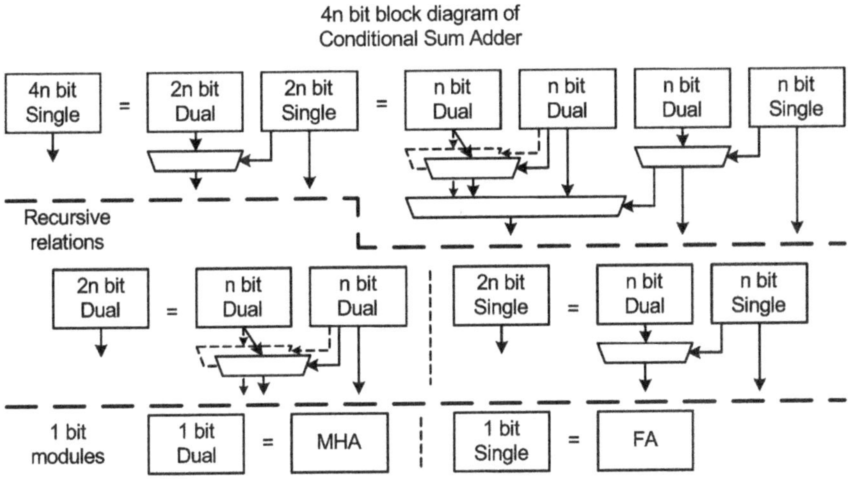

Figure 3.13 Recursive structure of a CSA. (© 2007 IEEE. *From* [1].)

$$s_i = s_i^0 \overline{c_i} + s_i^1 c_i \tag{3.16}$$

$$c_{i+1} = c_{i+1}^0 \overline{c_i} + c_{i+1}^1 c_i \tag{3.17}$$

As shown in Figure 3.14, the circuits are composed of a full adder (FA), modified half adders (MHAs), multiplexers (MUXs), and duplicated multiplexers (MUXDs). The schematics of these circuits are shown in Figures 3.15–3.17.

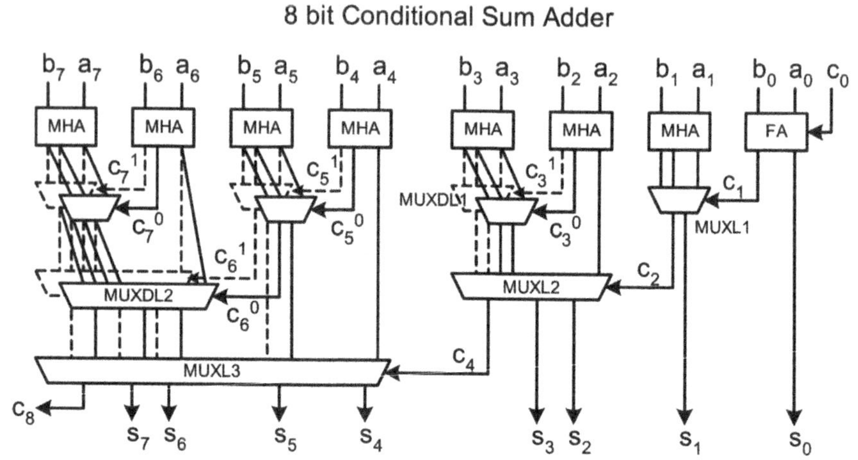

Figure 3.14 8-bit CSA block diagram. (© 2007 IEEE. *From* [1].)

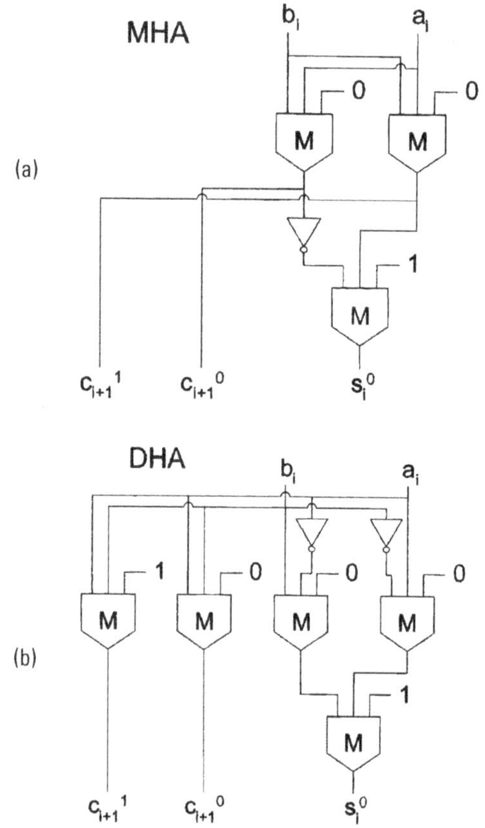

Figure 3.15 Half adder schematics: (a) MHA, (b) DHA. (© 2007 IEEE. *From* [1].)

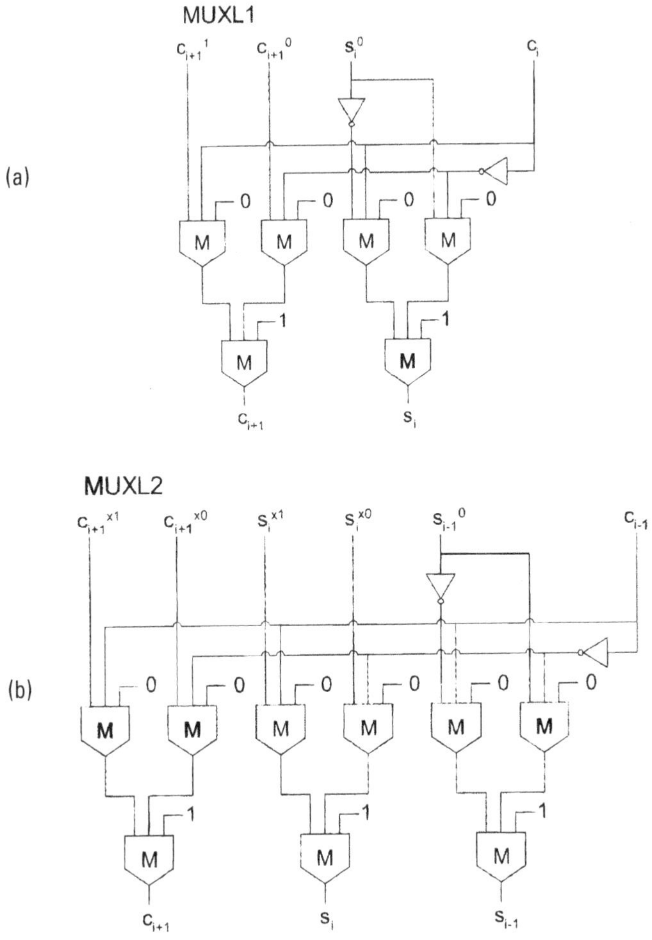

Figure 3.16 Level 1 and 2 MUX schematics. (a) Level-1 MUX. (b) Level-2 MUX. (© 2007 IEEE. *From* [1].)

Figure 3.15(a, b) show two options for the half adder modules. In transistor circuits, the duplicated half adder (DHA) has the same delay and more area than the MHA. In QCA, the area of the DHA is still large but the delay is slightly less than that of the MHA by one quarter of a clock. This trade-off governs the circuit choice. The general MHA module for CSA has four outputs, but the QCA design only has three outputs. The sum for a carry-in of one is the complement of the sum for a carry-in of zero. To reduce wire routing, the required inverter is made at the destination. Thus, two wire channels are reduced to one.

The MUXs shown in Figures 3.16 and 3.17 are similar to transistor circuits. The inverters for complementing the sums are implemented just before the MUXs. Successive levels of MUXs are implemented in the same manner.

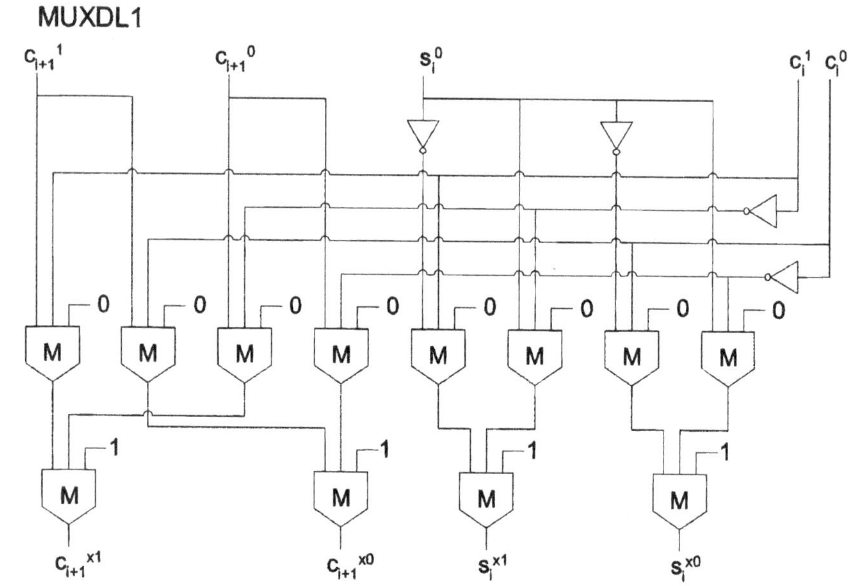

Figure 3.17 Level 1 duplicated MUX schematic. (© 2007 IEEE. *From* [1].)

3.4.3 Layout Design

Figure 3.18(a) and Figure 3.18(b) show the layouts for the two types of half adder modules. The DHA module has a larger area and more cells, but it has more delay margin. This shows a difference from transistor circuits. DHA modules are used in the implementations for the timing margin. The width of the design is dominated by the MUXs. The heights of the MHA and the DHA are the same. The only disadvantage of DHA is a small increase in the number of cells used, but this is negligible for large adders.

The layouts for 4- and 16-bit CSAs are shown in Figures 3.19 and 3.20.

3.5 Comparison of the Conventional Adders[4]

Table 3.2 compares the 4-, 8-, 16-, 32- and 64-bit RCAs, CLAs, and CSAs. Figures 3.21, 3.22, and 3.23 show comparison graphs of the three adders. From the statistics, cell counts for an adder with n-bit operands are roughly $O(n^{1.35})$ for RCAs, $O(n^{1.5})$ for CSAs. The areas are $O(n^{1.72})$for RCAs, $O(n^{1.47})$ for CLAs, and $O(n^{1.73})$for CSAs and the respective delays are $O(n^{0.97})$ for RCAs, $O(n^{0.8})$ for CLAs, and $O(n^{0.91})$ for CSAs. These results show that the design overheads are indeed significant.

4. Section 3.5 is based on [1].

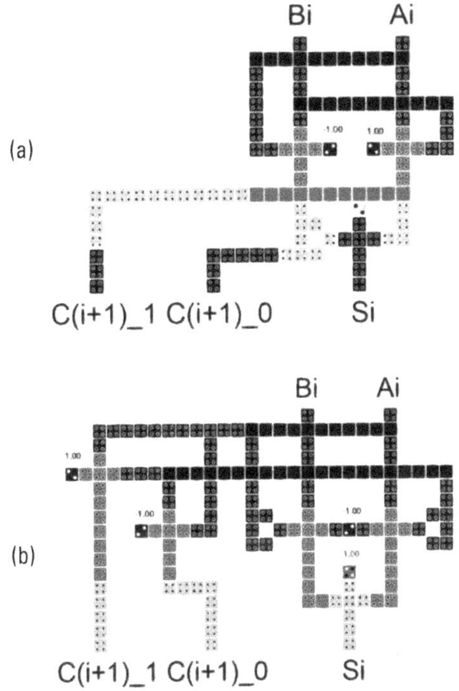

Figure 3.18 Half adder layout: (a) MHA, (b) DHA. (© 2007 IEEE. *From* [1].)

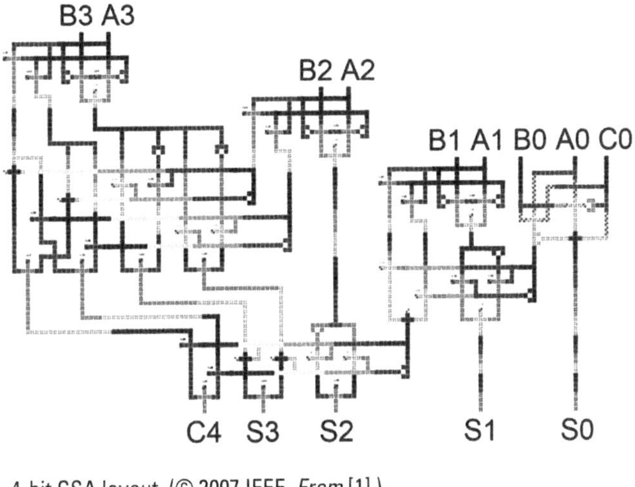

Figure 3.19 4-bit CSA layout. (© 2007 IEEE. *From* [1].)

Among the adders, the RCA is the simplest adder so its complexity is the lowest, but the dummy wire channels added for input/output synchronization require a large area. The CLA is the best conventional adder design in QCA.

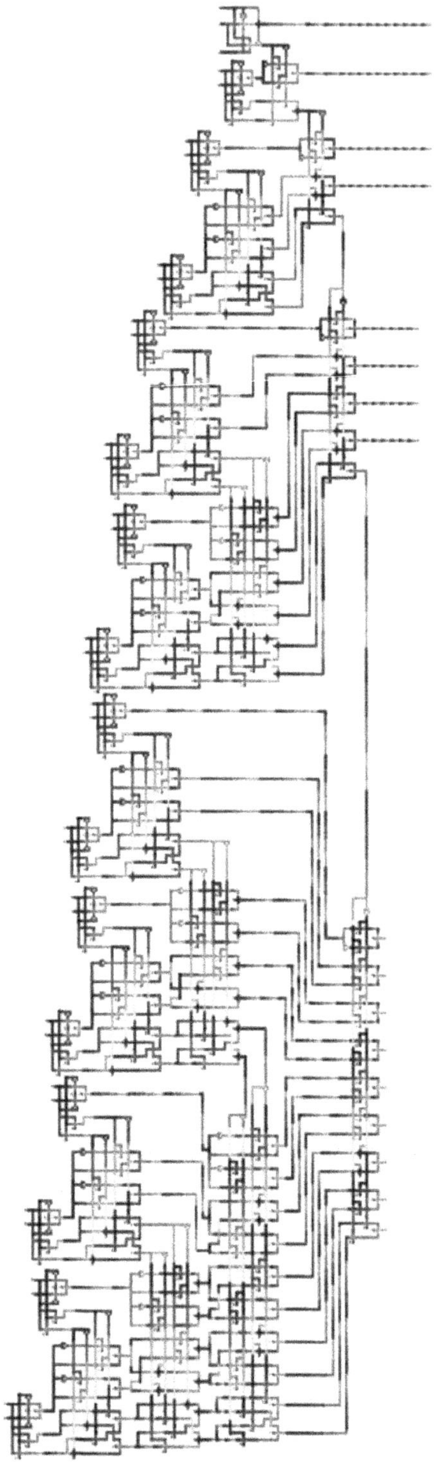

Figure 3.20 16-bit CSA layout. (© 2007 IEEE. *From* [1].)

Table 3.2
Adder Comparisons with Multilayer Crossovers

	Complexity	Area	Delay
RCA4	651 cells	1.67 x 0.72 μm^2	$4\frac{1}{4}$ clocks
RCA8	1,499 cells	3.43 X 1.04 μm^2	$8\frac{1}{4}$ clocks
RCA16	3,771 cells	6.97 x 1.69 μm^2	$16\frac{1}{4}$ clocks
RCA32	10,619 cells	14.03 x 3.01 μm^2	$32\frac{1}{4}$ clocks
RCA64	33,531 cells	28.18 x 5.65 μm^2	$64\frac{1}{4}$ clocks
CLA4	1,575 cells	1.74 x 1.09 μm^2	$3\frac{2}{4}$ clocks
CLA8	3,988 cells	3.50 x 1.58 μm^2	$6\frac{2}{4}$ clocks
CLA16	10,217 cells	7.02 x 2.21 μm^2	$10\frac{2}{4}$ clocks
CLA32	25,308 cells	14.06 x 3.05 μm^2	19 clocks
CLA64	59,030 cells	28.20 x 3.73 μm^2	$31\frac{2}{4}$ clocks
CSA4	1,999 cells	2.69 x 1.65 μm^2	$3\frac{3}{4}$ clocks
CSA8	6,216 cells	5.90 x 2.62 μm^2	$7\frac{3}{4}$ clocks
CSA16	16,866 cells	12.49 x 3.88 μm^2	14 clocks
CSA32	45,354 cells	25.67 x 6.17 μm^2	25 clocks
CSA64	129,611 cells	53.61 x 10.29 μm^2	45 clocks

The complexity is higher than that of the RCA, but the delay is less. The CSA shows the highest complexity and largest area due to the overhead from the MUXs and the wire channels. The areas reported in Table 3.2 are the size of the bounding box (i.e., the smallest rectangle that contains the layout). Since the design of the CSA is roughly triangular in shape, the size can be reduced by about 25% if the unused areas are taken into account, but it will still be the largest of the three types of adders.

The delays of the CLAs are less than that of the RCAs and CSAs when the operand size is large. Even though the area and complexity are increased in CLA, overall it is the best adder design in QCA. Comparing QCA circuits with transistor circuits, the main differences are observed in the CSAs. In transistor circuits, the CSA shows similar speed to the CLA, even though the size of the CSA is larger than that of the CLA. But in QCA, the CSA is slower and the complexity and area are much greater than those of the CLA.

Generally, QCA circuits have very significant wiring delays. For a fast design in QCA, complexity constraints are very critical issues and the design needs to use architectural techniques to boost the speed considering these limitations. Section 3.6 takes a closer look at the RCA to see if an improved FA can produce better performance.

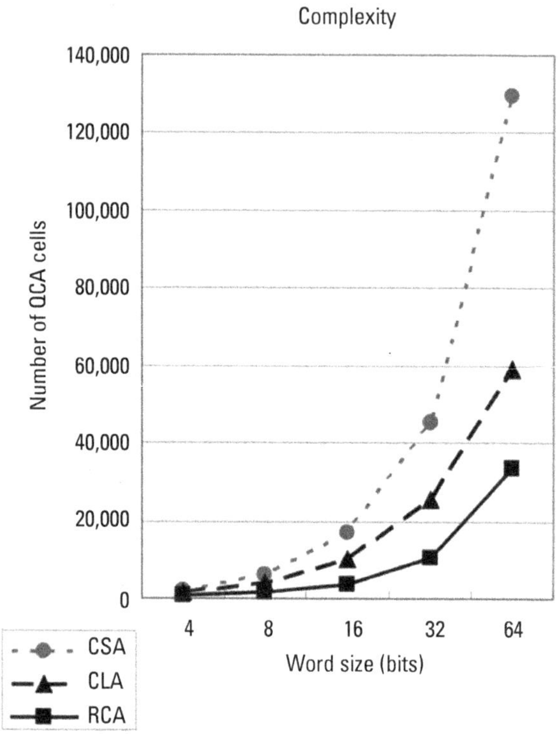

Figure 3.21 Complexity comparisons of the "conventional" adders. (© 2007 IEEE. *From* [1].)

3.6 Carry Flow Adder[5]

3.6.1 Basic Design Approach

The previous sections show that interconnections incur significant complexity and wire delays when implemented in QCA, so transistor circuit designs that assume wires have negligible complexity and delay need to be re-examined. In QCA, if the complexity increases, the delay may increase because of the increased cell counts and wire connections. In this section, the adder design is that of a conventional RCA, but with a FA whose layout is optimized for QCA technology. The proposed adder design shows that a very low delay can be obtained with an optimized layout. This is in contrast to the conventional RCA. To avoid confusion with conventional RCAs such as those described in Section 3.2, the new layout is referred to as the CFA.

Equations for a FA realized with majority gates and inverters are shown below. Most adder delays come from carry propagation. For faster calculation, reducing the carry propagation delay is most important. The usual approach

5. Section 3.6 is based on [8].

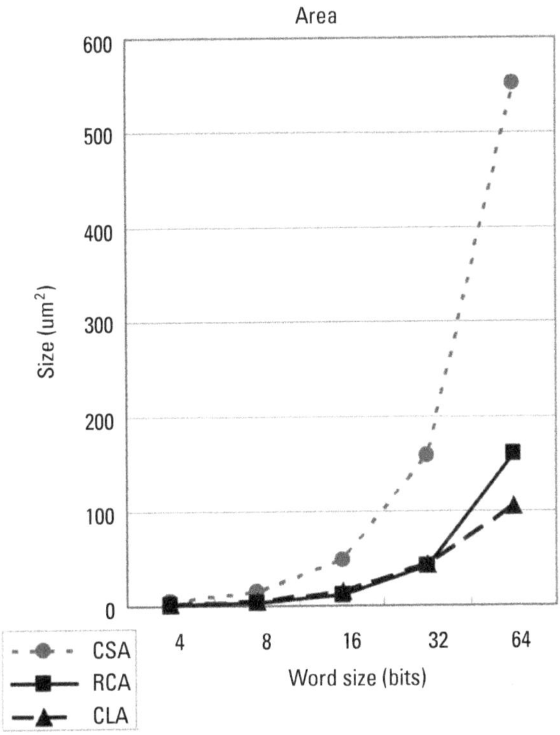

Figure 3.22 Area comparison of the "conventional" adders. (© 2007 IEEE. *From* [1].)

for fast carry propagation is to add additional logic elements. In this design, simplification is used instead.

In QCA, the path from carry-in to carry-out only uses one majority gate. The majority gate always adds one more clock zone (one quarter clock delay). Thus, each bit in the words to be added requires at least one clock zone, which sets the minimum delay.

$$s_i = a_i b_i c_i + a_i \overline{b_i} \overline{c_i} + \overline{a_i} b_i \overline{c_i} + \overline{a_i} \overline{b_i} c_i \tag{3.18}$$

$$s_i = M\left(\overline{M}\left(a_i, b_i, c_i\right), M\left(a_i, b_i, \overline{c_i}\right) c_i\right) \tag{3.19}$$

$$s_i = M\left(\overline{c}_{i+1} M\left(a_i, b_i, \overline{c_i}\right) c_i\right) \tag{3.20}$$

$$c_{i+1} = a_i b_i + b_i c_i + a_i c_i \tag{3.21}$$

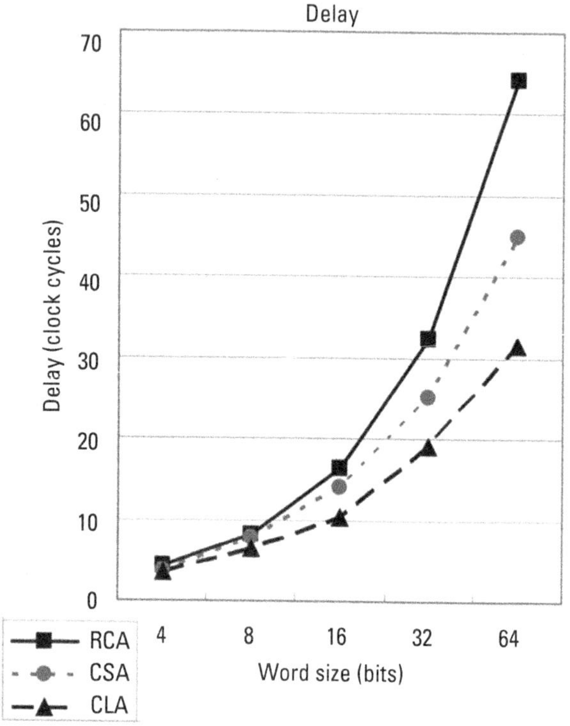

Figure 3.23 Delay comparison of the "conventional" adders. (© 2007 IEEE. *From* [1].)

$$c_{i+1} = M\left(a_i, b_i, c_i\right) \tag{3.22}$$

3.6.2 Carry Flow Full Adder Design

Based on previous approaches, a 1-bit FA is designed. The input bit streams flow downward, and the carry propagates from right to left. Figure 3.24(a, b) show the schematic and the layout of the carry flow FA. The schematic and layout are optimized to minimize the delay and area. The carry propagation delay for 1-bit is a quarter clock, and the delay from data inputs to the sum output is three quarter clocks.

The wiring channels for the input/output synchronization should be minimized since wire channels add significantly to the circuit area. The carry flow FA shown in Figure 3.24(b) requires a vertical offset between the carry-in and carry-out of only one cell.

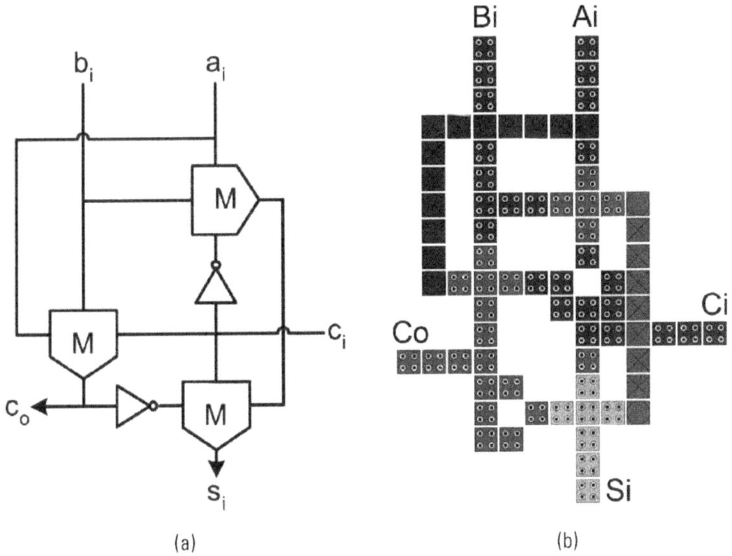

Figure 3.24 1-bit FA for the CFA: (a) schematic, (b) layout. (© 2009 IEEE. *From* [8].)

Figures 3.25 and 3.26 show 4-bit and 32-bit RCAs, respectively, realized with carry flow FAs. From the layouts, it is clear that for large adders, much of the area is devoted to skewing the input data and deskewing the outputs.

3.6.3 Simulation Results

For clarity, only 8-bit CFA simulation results are shown. The input and output waveforms are shown in Figure 3.27. The first meaningful output appears in

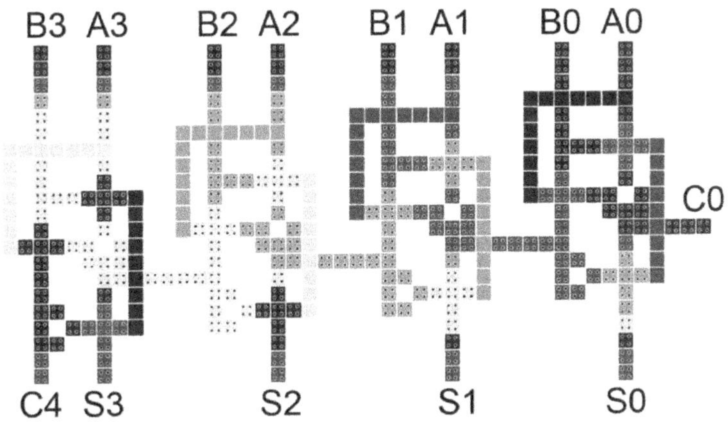

Figure 3.25 Layout of a 4-bit CFA. (© 2009 IEEE. *From* [8].)

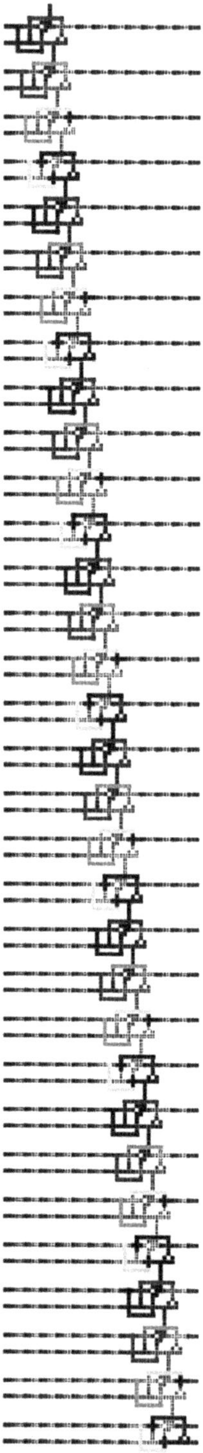

Figure 3.26 Layout of a 32-bit CFA. (© 2009 IEEE. *From* [8].)

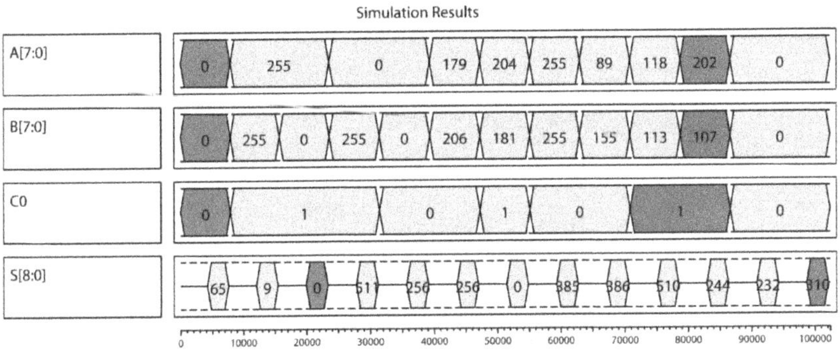

Figure 3.27 Simulation results for 8-bit CFA. (© 2009 IEEE. *From* [8].)

the third clock period after $2\frac{2}{4}$ clock delays. First and last input/output pairs are highlighted.

For design comparisons, QCA CLAs (CLA) are used since they were shown to be smaller and faster than conventional RCAs and CSAs in Sections 3.2–3.4. Table 3.3 shows comparisons of the 4-, 8-, 16-, 32-, and 64-bit designs for the CLA and CFA. Figures 3.28–3.30 compare the two types of adders.

From the statistics, the cell count for the CFA with n-bit operands is roughly $O(n^{1.21})$, the area is $O(n^{1.42})$and the delay for the CFA-based RCA is proportional to the word size after a half clock start up delay. From the comparison with the CLA, the complexity, area, and delay are much better with the CFA FA, so with a carefully optimized FA design a QCA RCA is faster and smaller than a CLA.

3.7 Decimal Adder[6]

One of the earliest electronic computers, the ENIAC, was designed using decimal arithmetic [10]. However, with the development of computer architecture, binary arithmetic became the standard number system for electronic computers, and it now dominates the modern computing world. While binary computer arithmetic design has been extensively investigated [11–13], limited attention has been given to decimal arithmetic. However, new financial, commercial and internet-based applications demand high-accuracy decimal floating point computations that cannot be achieved with binary arithmetic. The importance of decimal arithmetic has been recognized, and its specification has been included in the recent revision to the IEEE 754 standard [14]. Hence, this section explores the possibility of implementing decimal arithmetic in QCA technology.

6. Section 3.7 is based on [9].

Table 3.3
Adder Comparisons Between CLA and CFA

	Complexity	Area	Delay
CLA4	1,575 cells	1.74 x 1.09 μm^2	$3\frac{2}{4}$ clocks
CLA8	3,988 cells	3.50 x 1.58 μm^2	$6\frac{2}{4}$ clocks
CLA16	10,217 cells	7.02 x 2.21 μm^2	$10\frac{2}{4}$ clocks
CLA32	25,308 cells	14.06 x 3.05 μm^2	19 clocks
CLA64	59,030 cells	28.20 x 3.73 μm^2	$31\frac{2}{4}$ clocks
CFA4	371 cells	0.90 x 0.45 μm^2	$1\frac{2}{4}$ clocks
CFA8	789 cells	1.79 x 0.53 μm^2	$2\frac{2}{4}$ clocks
CFA16	1,769 cells	3.55 x 0.69 μm^2	$4\frac{2}{4}$ clocks
CFA32	4,305 cells	7.09 x 1.03 μm^2	$8\frac{2}{4}$ clocks
CFA64	11,681 cells	14.15 x 1.71 μm^2	$16\frac{2}{4}$ clocks

Decimal adders are the core of decimal arithmetic and some previous work has been conducted into QCA decimal adders [15–17]. In [15], the authors presented a conventional BCD adder using RCAs (RCAs). Since the layout of the RCA that was used is not optimized, the BCD adder occupies a large area and introduces long delay. In [16], the majority implementation approach (MIA) is used to reduce the correction logic. However, the layout of this design can be further optimized to reduce the area cost. Parallel and serial QCA decimal adders using the Johnson-Mobius code (JMC) were designed in [17].

The BCD 8,4,2,1 number system is used here as it is widely used for decimal arithmetic. Two types of BCD adders are proposed: one is a CFA-based BCD adder with correction logic and the other is a CLA-based BCD adder . Compared with the previous QCA designs, both designs achieve better performance in terms of latency and overall cost.

3.7.1 Conventional BCD Adder

In BCD addition, each BCD digit of the sum is the sum of two 4-bit binary numbers. When the binary sum is less than $(10)_{10}$ the corresponding BCD digit sum is correct. However, when the binary sum is greater than or equal to $(10)_{10}$, correction logic is required to get the right result. An addition of $(6)_{10}$ to the binary sum is used for conversion. The correction logic is designed by detecting the occurrence of the binary numbers from $(1010)_2$ to $(10011)_2$. An outline of one-digit BCD addition is shown in Figure 3.31, where A = $a_3a_2a_1a_0$ and $B = b_3b_2b_1b_0$ are two BCD inputs, c_4, c_3, c_2, c_1 are the binary carry outs from the lower order binary additions, c_{in} and c_{out} are the decimal carry in and carry out respectively, $Z = z_3z_2z_1z_0$ is the binary sum and $S = s_3s_2s_1s_0$ is the final BCD sum.

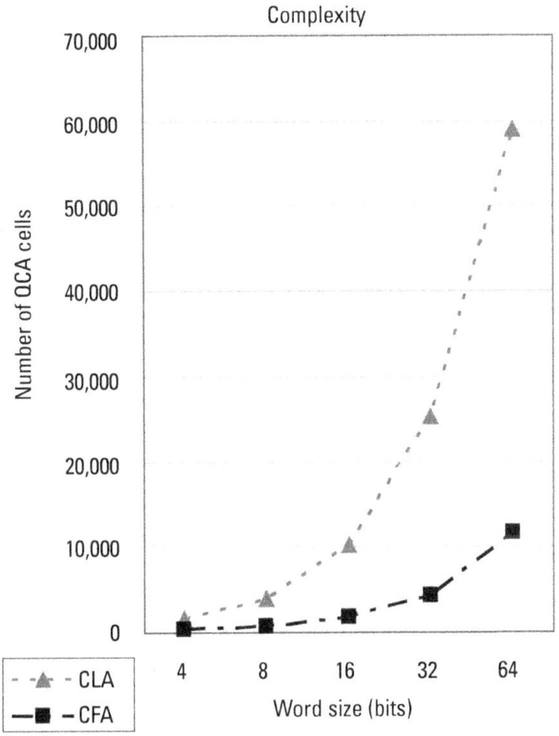

Figure 3.28 Complexity comparison of CFA and CLA adders for various operand sizes. (© 2009 IEEE. *From* [8].)

3.7.1.1 Schematic Design

For a conventional BCD adder, the final BCD sum can be calculated as follows (the bits shown in bold type do not affect the sum, so they can be ignored in the addition): if $S \geq (10)_{10}$ ($c_{out} = 1$), then

$$S = Z + (6)_{10} = Z + (0110)_2 \tag{3.23}$$

if $S < (10)_{10}$ ($c_{out} = 0$), then

$$S = Z + (0)_{10} = Z + (0000)_2 \tag{3.24}$$

The carry out can be expressed with majority logic as follows:

$$c_{out} = c_4 + (z_1 + z_2)z_3 = M\left\{c_4, 1, M\left[\left(z_3, 0, M(z_1, 1, z_2)\right)\right]\right\} \tag{3.25}$$

Eight binary adders are usually required for a one-digit BCD adder. Four adders are used for the binary sum and another four adders are used for the

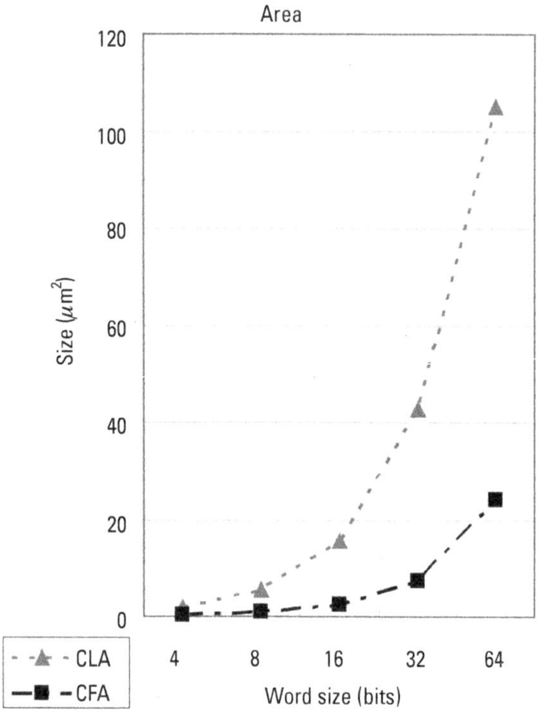

Figure 3.29 Area comparison of CFA and CLA adders for various operand sizes. (© 2009 IEEE. *From* [8].)

sum conversion. As the correction step only adds $(6)_{10}$ or $(0)_{10}$ to the binary sum, the least significant bit of the decimal sum s_0 is always equal to z_0, (i.e., $s_0 = z_0$). Based on this observation [18], one binary adder can be removed in the correction step. The design performance of a conventional BCD adder heavily depends on the binary adder architectures. Therefore, the CFA [8], the most efficient QCA binary adder to date, is used in the proposed conventional BCD adder design. A block diagram of the CFA-based BCD adder is shown in Figure 3.32.

3.7.1.2 Layout Design and Simulation

The layout of a one-digit CFA-based BCD adder is shown in Figure 3.33. Multilayer crossovers are used in this work. The design has been verified by QCADesigner [6] and the simulation results of the CFA-based BCD adder are shown in Figure 3.34. In the simulations, most default parameters for a bistable approximation as listed in Section 2.4.1.3 are used. The number of samples is determined to be 128,000.

From the simulation results, it can be seen that the latency of the CFA-based BCD adder is 4.75 clock cycles. A detailed comparison between this CFA-based BCD adder and previous designs is provided in Section 3.7.3.

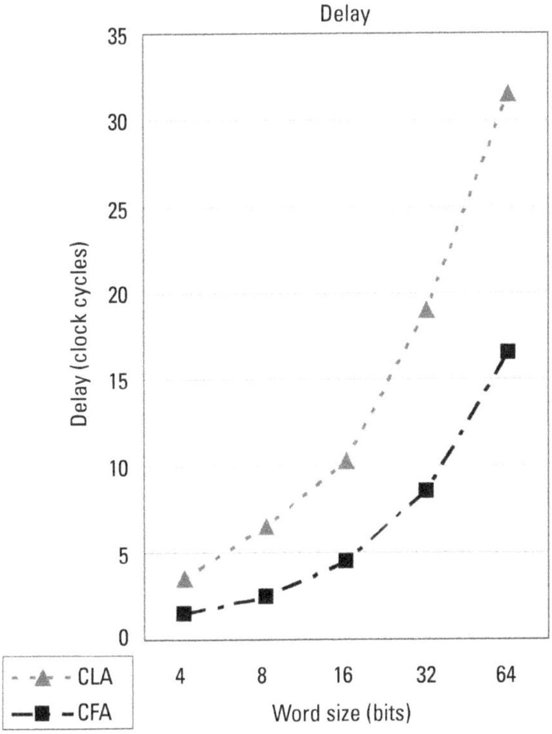

Figure 3.30 Delay comparison of CFA and CLA adders for various operand sizes. (© 2009 IEEE. *From* [8].)

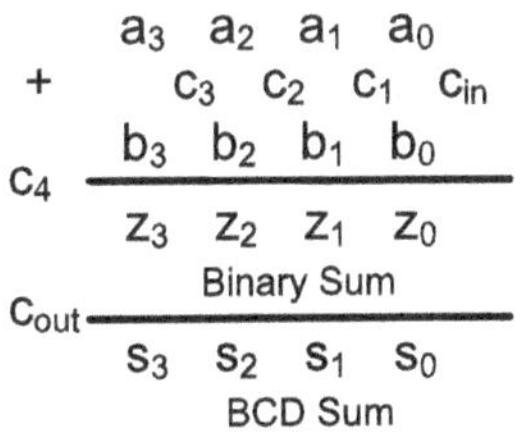

Figure 3.31 One-digit BCD addition. (© 2012 IEEE. *From* [9].)

3.7.2 Carry Lookahead Decimal Adder

Conventional BCD adders have significant latency due to the correction operations and the usage of slow RCAs. To achieve high speed decimal addition, an alternative method was investigated. In this section, a CLA-based BCD adder is proposed using majority optimization in QCA technology.

3.7.2.1 Schematic Design for a Carry Lookahead BCD Adder

Carry generation g_i and carry propagation p_i are defined as follows:

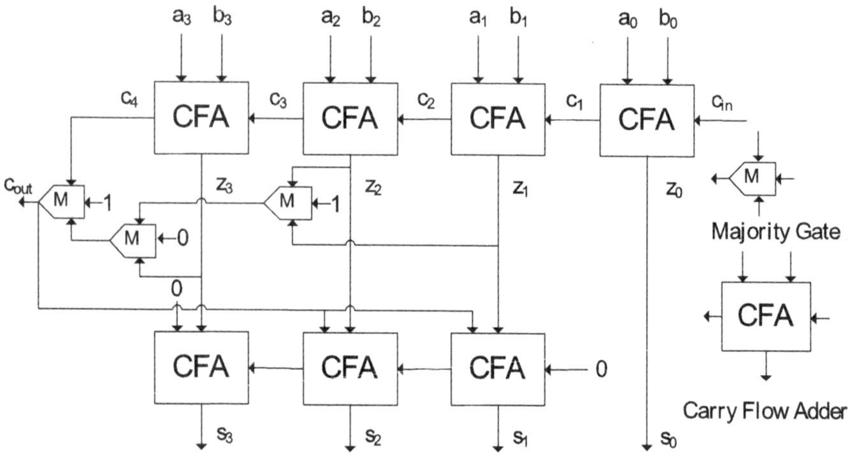

Figure 3.32 Block diagram of the CFA-based BCD adder. (© 2012 IEEE. *From* [9].)

$$g_i = a_i \cdot b_i \tag{3.26}$$

$$p_i = a_i + b_i \tag{3.27}$$

Since $p_i g_i = g_i$ and $p_i + g_i = p_i$, then $g_i + p_i X = M(g_i, pi, X)$, where X can be any logic function [19]. Furthermore, the c_i and z_i can be expressed using p_i and g_i as follows:

$$c_{i+1} = g_i + p_i c_i = M\left(g_i, p_i, c_i\right) \tag{3.28}$$

$$z_i = M\left(\overline{c}_{i+1}, M\left(g_i, p_i, \overline{c}_{i+1}\right), c_i\right) \tag{3.29}$$

The decimal sum can be obtained by observing all input combinations that generate '1' at each bit of the final BCD sum [20]. The derivation process of the decimal sum can be divided into two cases: Is the sum greater than or equal to $(10)_{10}$, or is the sum less than $(10)_{10}$.

1. Case I: Sum ≥10. In this case: s_3 is '1' if the sum is 18 =$(0001, 1000)_{BCD}$ or 19 = $(0001, 1001)_{BCD}$; s_2 is '1' if the sum is 14 = $(0001, 0100)_{BCD}$, 15 = $(0001, 0101)_{BCD}$, 16 = $(0001, 0110)_{BCD}$ or 17 = $(0001, 0111)_{BCD}$; $s_1 = \overline{z}_1$ and $s_0 = z_0$. Therefore, the bits of the BCD sum can be expressed using p_i, g_i and majority gates as follows, when the sum ≥ $(10)_{10}$:

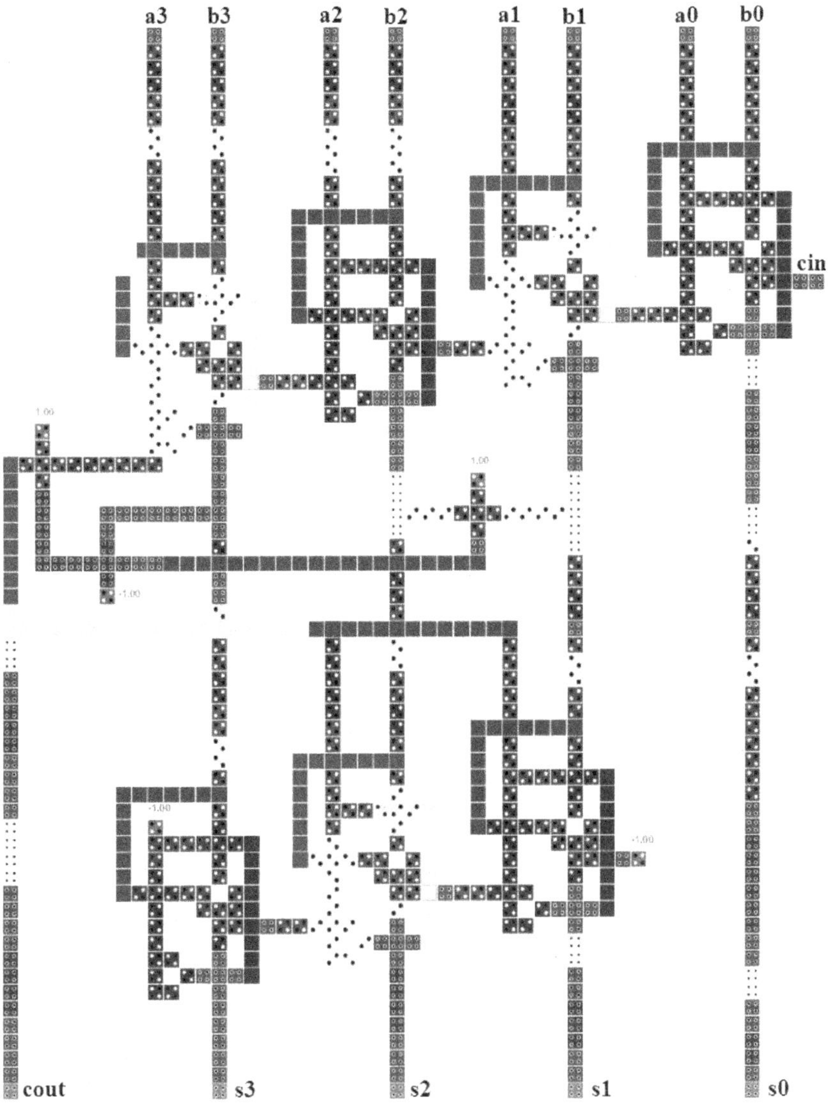

Figure 3.33 Layout of one-digit CFA-based BCD adder. (© 2012 IEEE. *From* [9].)

$$s_{3-I} = g_3 c_1 = g_3 M\left(g_0, p_0, c_{\text{in}}\right) \tag{3.30}$$

$$s_{2-I} = g_3 \overline{c_1} + p_3 p_2 \left(p_1 + c_1\right) + g_2 g_1 c_1 \tag{3.31}$$

$$s_{1-I} = \overline{z_1} = \overline{M}\left(\overline{c_2}, M\left(g_1, p_1, \overline{c_2}\right), c_1\right) \tag{3.32}$$

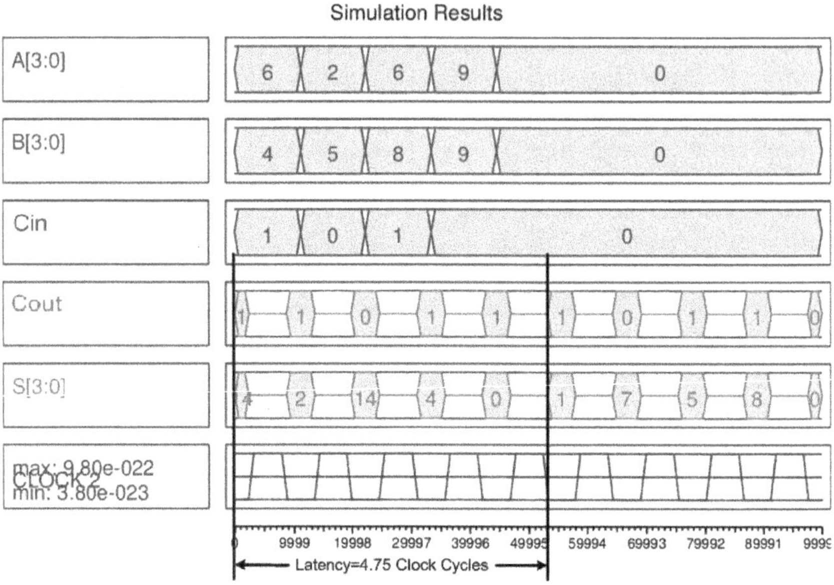

Figure 3.34 Simulation results of one-digit CFA-based BCD adder. (© 2012 IEEE. *From* [9].)

$$s_{0-I} = z_0 = M\left(\overline{c}_1, M\left(g_0, p_0, \overline{c}_1\right), c_{\text{in}}\right) \tag{3.33}$$

2. Case II: Sum < 10: In this case: s_3 is '1' if the sum is 8 $=(1000)_{\text{BCD}}$ or 9 $= (1001)_{\text{BCD}}$; s_2 is '1' if the sum is 4 $= (0100)_{\text{BCD}}$, 5 $= (0101)_{\text{BCD}}$, 6 $=(0110)_{\text{BCD}}$ or 7 $=(0111)_{\text{BCD}}$; $s_1 = z_1$ and $s_0 = z_0$. Therefore, the bits of the BCD sum can be expressed as follows, when the sum $< (10)_{10}$:

$$s_{3-II} = p_3 + c_3 = p_3 + M\left(g_2, p_2, c_2\right) \tag{3.34}$$

$$s_{2-II} = \overline{c}_3\left(p_2 + c_2\right) = \overline{M}\left(g_2, p_2, c_2\right)\left(p_2 + c_2\right) \tag{3.35}$$

$$s_{1-II} = z_1 = M\left(\overline{c}_2, M\left(g_1, p_1, \overline{c}_2\right), c_1\right) \tag{3.36}$$

$$s_{0-II} = z_0 = M\left(\overline{c}_1, M\left(g_0, p_0, \overline{c}_1\right), c_{\text{in}}\right) \tag{3.37}$$

3. Carry-out generation: The direct BCD result depends on the decimal carry-out. In this work, the decimal carry-out proposed in [21] is used

$$c_{out} = g_3 + p_3 p_2 + (p_3 + g_2) p_1 + [p_3 + g_2 + p_2 g_1] c_1 \tag{3.38}$$

and can be optimized with majority logic as follows:

$$c_{out} = M(g_3, p_3, p_2) + (p_3 + g_2) p_1 + [p_3 + M(g_2, p_2, g_1)] c_1 \tag{3.39}$$

The selection of the sum bits for the different cases is performed by three 2:1 MUXs. A block diagram of the CLA-based BCD adder is shown in Figure 3.35.

3.7.2.2 Layout Design and Simulation

The layout of a one-digit CLA-based BCD adder is shown in Figure 3.36. The cell size is the same as that of Figure 3.33, so the areas of the adders can be compared easily. The simulation results of the adder from QCADesigner are shown in Figure 3.37. From the simulation results, it can be seen that the latency of the CLA decimal adder is only 2.5 clock cycles. A detailed comparison between the CLA-based BCD adder and previous designs is provided in Section 3.7.3.

3.7.3 Comparison and Analysis

Table 3.4 provides a comparison of the two proposed BCD adders with all previously published QCA decimal adders. The proposed and published designs are all one-digit designs. The decimal adders are compared in terms of cell

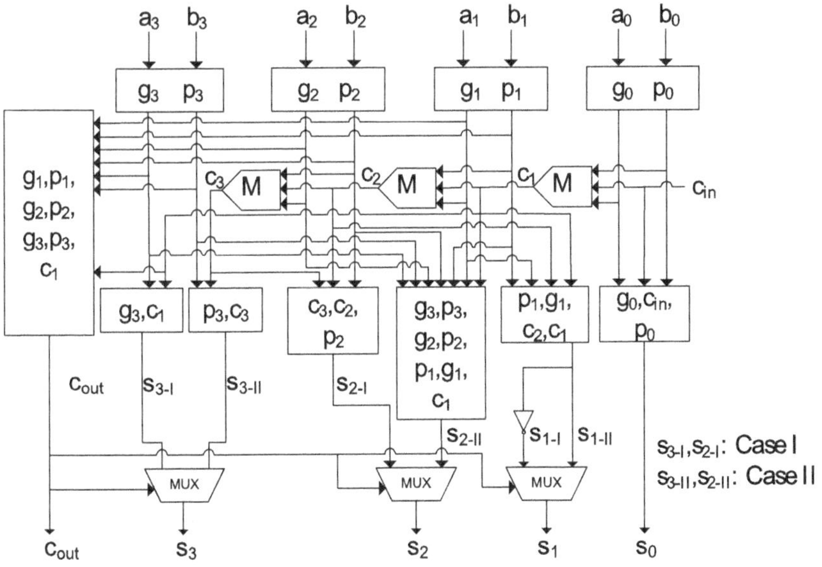

Figure 3.35 Block diagram of the proposed CLA-based BCD adder. (© 2012 IEEE. *From* [9].)

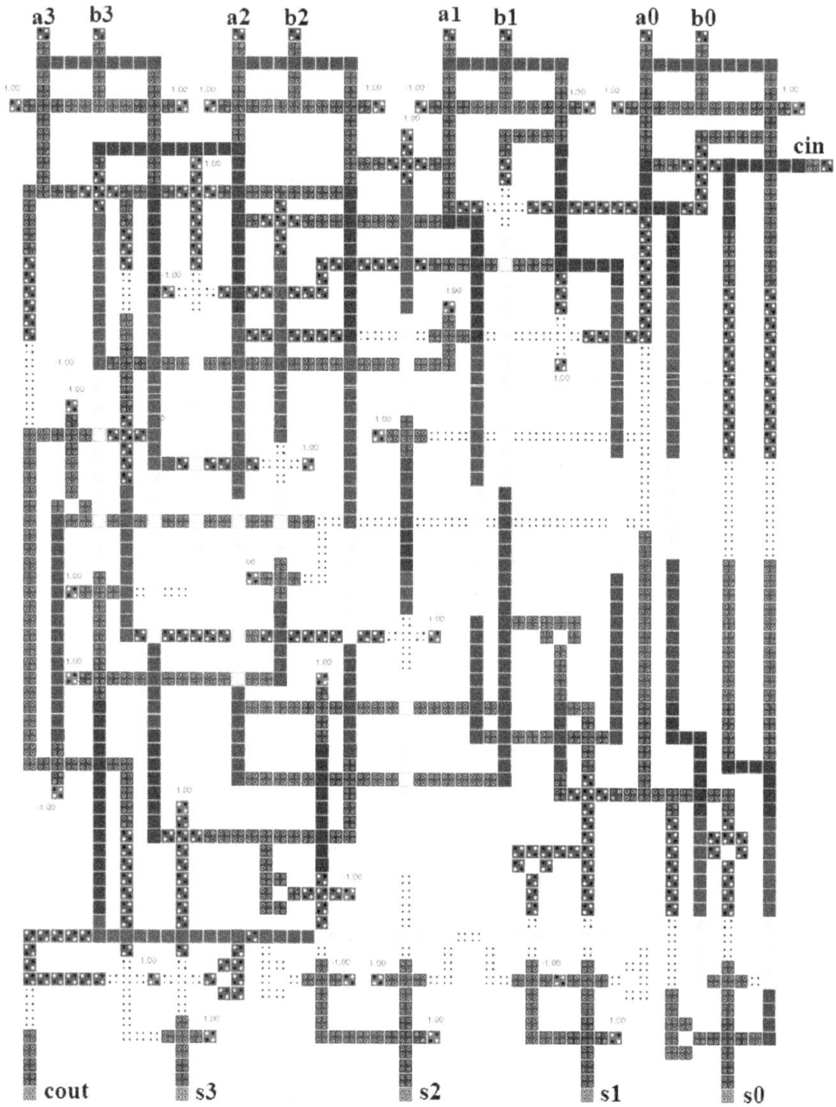

Figure 3.36 Layout of a one-digit CLA-based BCD adder. (© 2012 IEEE. *From* [9].)

count, area, latency, and overall cost. All designs are compared with the same cell size (18 nm × 18 nm). The overall cost function is defined as follows [22]:

$$\text{Overall Cost} = \text{Area} \times \text{Latency}^2 \tag{3.40}$$

It is clear from Table 3.4 that the proposed CFA-based BCD adder is the smallest QCA BCD adder due to optimizing the correction logic, which saves

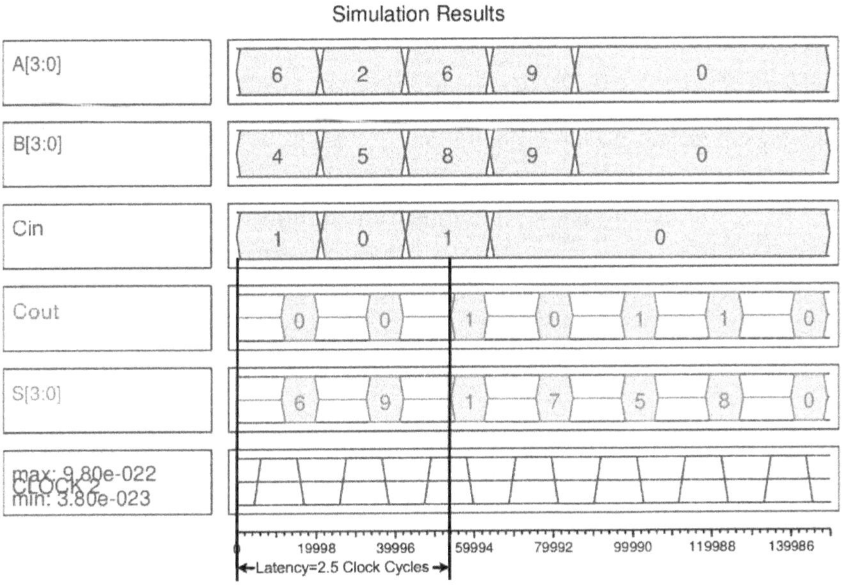

Figure 3.37 Simulation results of one digit CLA-based BCD adder. (© 2012 IEEE. *From* [9].)

Table 3.4
QCA Decimal Adder Comparison

QCA Decimal Adders (One-Digit)	**Cell Count**	**Area (μm^2)**	**Latency (Number Cycles)**	**Overall Cost**
Parallel BCD Adder [15]	NA	5.80	NA	NA
Parallel BCD Adder [16]	1,348	2.28	8	146
Parallel JMC Adder [17]	3,560	4.00	7	196
Serial JMC Adder [17]	1,130	1.77	10	177
CFA-based BCD Adder	932	1.36	4.75	31
CLA-based BCD Adder	1,838	1.86	2.5	12

a binary adder. It also achieves a faster speed than all previous designs with a latency of 4.75 clock cycles and it achieves a lower overall cost with a reduction of more than 80% over the previous most cost-efficient decimal adder (i.e., the serial JMC adder). The proposed CLA-based BCD adder is even faster with a latency of only 2.5 clock cycles, an increase in speed of over 60% when compared with the previous fastest decimal adder (i.e., the parallel JMC adder). This is as a result of directly calculating the decimal sum. Furthermore, its overall cost is the lowest, over 90% less than the serial JMC adder.

3.8 Conclusion

Generally QCA circuits have very significant wiring delays. For a fast design in QCA, complexity constraints are very critical issues and the design needs to use architectural techniques to boost the speed considering these limitations. This chapter presents designs for practical QCA adders.

RCAs, CLAs, and CSAs are designed and optimized for QCA technology. The delays of the proposed CLAs are much less than the RCAs and CSAs when the operand size is large. Even though the complexity is higher for CLA than RCA, overall it is the best adder design in QCA. Comparing QCA circuits with transistor circuits, the main differences are observed in the CSAs. In transistor circuits, the CSA shows similar speed to the CLA, even though the size of the CSA is larger than that of CLA. But in QCA, the CSA is slower and the complexity and area are much more than those of the CLA.

This chapter also presents a new adder design, the CFA. CFAs use conventional carry propagation schemes but are optimized for layout in QCA technology. Compared with other adder designs, the CFA shows the smallest complexity, area, and delay. So the CFA is a fast and efficient adder in QCA.

Two cost-efficient QCA BCD adder designs are proposed in this chapter. A CFA-based BCD adder is designed using the most efficient binary adder, the CFA. Its correction logic is optimized by saving a binary adder during the correction step. An even faster CLA-based decimal adder is designed by directly calculating the decimal sum. Compared with previous designs, both decimal adders achieve better performance in terms of latency and overall cost. The proposed CFA-based BCD adder has the smallest area with the least number of cells. The proposed CLA-based BCD adder is the fastest with an increase in speed of over 60% when compared with the previous fastest decimal QCA adder. It also has the lowest overall cost with a reduction of over 90% when compared with the previous most cost-efficient design.

References

[1] Cho, H., and E. Swartzlander, Jr., "Adder Designs and Analyses for Quantum-Dot Cellular Automata," *IEEE Transactions on Nanotechnology,* Vol. 6, 2007, pp. 374–383.

[2] Vetteth, A., et al., "Quantum-Dot Cellular Automata Carry-Lookahead Adder and Barrel Shifter," in *Proceedings of the IEEE Emerging Telecommunications Technologies Conference,* 2002, pp. 1–5.

[3] Wang, W., K. Walus, and G. Jullien, "Quantum-Dot Cellular Automata Adders," in *Proceedings of the 3rd IEEE Conference on Nanotechnology,* Vol. 1, 2003, pp. 461–464.

[4] Cho, H., and E. Swartzlander, Jr., "Pipelined Carry Lookahead Adder Design in Quantum-dot Cellular Automata," in *Conference Record of the 39th Asilomar Conference on Signals, Systems and Computers,* 2005, pp. 1191–1195.

[5] Unwala, I. H., and E. E. Swartzlander Jr, "Superpipelined adder designs," in *Proceedings of the IEEE International Symposium o Circuits and Systems,* 1993, pp. 1841–1844.

[6] Walus, K., "QCADesigner," website, 2013, http://www.mina.ubc.ca/qcadesigner.

[7] Cho, H., and E. Swartzlander, Jr., "Modular Design of Conditional Sum Adders Using Quantum-Dot Cellular Automata," in *Proceedings of the 6th IEEE Conference on Nanotechnology,* Vol. 1, 2006, pp. 363–366.

[8] Cho, H., and E. Swartzlander, Jr., "Adder and Multiplier Design in Quantum-Dot Cellular Automata," *IEEE Transactions on Computers,* Vol. 58, 2009, pp. 721–727.

[9] Liu, W., et al., "Cost-Efficient Decimal Adder Design in Quantum-Dot Cellular Automata," in *Proceedings of the IEEE International Symposium on Circuits and Systems,* 2012, pp. 1347–1350.

[10] Aspray, W., et al., *Computing Before Computers,* Ames, Iowa: Iowa State University Press, 1990.

[11] Walus, K., G. Jullien, and V. Dimitrov, "Computer Arithmetic Structures for Quantum Cellular Automata," in *Conference Record of the 37th Asilomar Conference on Signals, Systems and Computers,* Vol. 2, 2003, pp. 1435–1439.

[12] Hänninen, I., and J. Takala, "Arithmetic Design on Quantum-Dot Cellular Automata Nanotechnology," in *Proceedings of the 8th International Workshop on Embedded Computer Systems: Architectures, Modeling, and Simulation,* 2008, pp. 43–52.

[13] Swartzlander, Jr., E., et al., "Computer Arithmetic Implemented with QCA: A Progress Report," in *Conference Record of the 44th Asilomar Conference on Signals, Systems and Computers,* 2010, pp. 1392–1398.

[14] "IEEE Standard for Floating-Point Arithmetic," *IEEE Std. 754-2008,* 2008. [Online]. Available: http://grouper.ieee.org/groups/754/

[15] Taghizadeh, M., M. Askari, and K. Fardad, "BCD Computing Structures in Quantum-dot Cellular Automata," in *Proceedings of the IEEE International Conference on Computer and Communication Engineering,* 2008, pp. 1042–1045.

[16] Kharbash, F., and G. Chaudhry, "The Design of Quantum-Dot Cellular Automata Decimal Adder," in *Proceedings of the IEEE International Multitopic Conference,* 2008, pp. 71–75.

[17] Gladshtein, M., "Quantum-Dot Cellular Automata Serial Decimal Adder," *IEEE Transactions on Nanotechnology,* Vol. 10, 2011, pp. 1377–1382.

[18] Richards, R., *Arithmetic Operations in Digital Computers,* New York: D. Van Nostrand Co., 1955.

[19] Pudi, V., and K. Sridharan, "Low Complexity Design of Ripple Carry Adder and Brent-Kung Adders in QCA," *IEEE Transactions on Nanotechnology,* Vol. 11, 2012, pp. 105–119.

[20] You, Y., Y. Kim, and J. Choi, "Dynamic Decimal Adder Circuit Design by Using the Carry Lookahead Adder," in *Proceedings of the IEEE Symposium on Design and Diagnostics of Electronic Circuits and Systems,* 2006, pp. 242–244.

[21] Schmookler, M., and A. Weinberger, "High Speed Decimal Addition," *IEEE Transactions on Computers,* Vol. 100, 1971, pp. 862–866.

[22] Thompson, C. D., "Area-Time Complexity for VLSI," in *Proceedings of the 7th Annual ACM Symposium on Theory of Computing,* 1979, pp. 81–88.

4

QCA Multipliers

Seong-Wan Kim and Earl E. Swartzlander, Jr.

Many digital signal processing systems use multiplication. It is an important operation in ALUs. The speed of multiplication limits the operation rate of the system for most applications. Parallel multipliers provide fast multiplication, but they need a large number of transistors, resulting in high power consumption and an expensive cost in CMOS. Many algorithms, architectures, and technologies have been proposed to improve the function of multipliers [2–5].

This chapter explores the implementation of three types of parallel multipliers in QCA technology. Array multipliers considered in Section 4.2 are well-suited to QCA implementation since they are formed by a regular array of identical functional units. The structure is conformable to QCA technology without extra wire delay. Column compression multipliers, such as Wallace and Dadda, multipliers are implemented with several different operand sizes in Section 4.3. Finally, Section 4.4 describes quasi-modular parallel multipliers that facilitate the use of irregular tree structure multipliers to make them more suitable for QCA. A detailed analysis of all the designs is discussed in Section 4.5. Section 4.6 provides a summary of multiplier designs in QCA.

4.1 Introduction[1]

This section presents the basics of binary multiplication. Figure 4.1 shows the multiplication of two n-bit unsigned binary numbers that yields a $2n$-bit product.

1. Chapter 4 is based on [1].

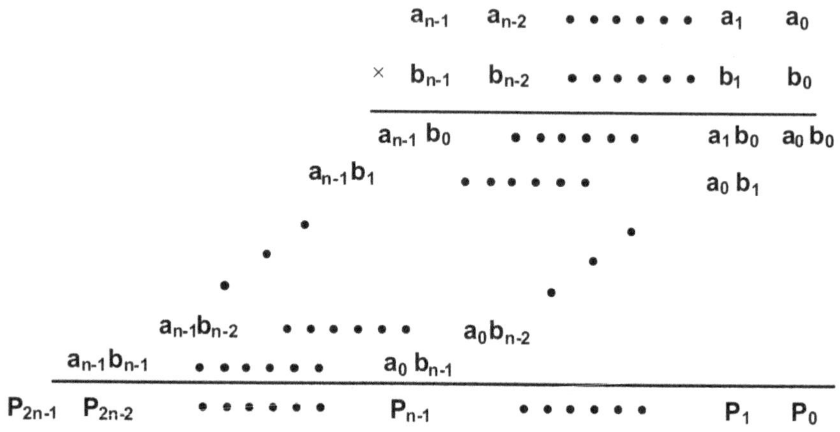

Figure 4.1 Multiplication of two n-bit binary numbers.

The basic equation is:

$$\begin{aligned} P &= A \times B \\ &= \sum_{j=0}^{n-1} a_j 2^j \times \sum_{i=0}^{n-1} b_i 2^i \\ &= \sum_{k=0}^{2n-1} p_k 2^k \end{aligned} \tag{4.1}$$

where, the multiplicand is $A(= a_{n-1}, \ldots, a_0)$, the multiplier is $B(= b_{n-1}, \ldots, b_0)$, and the product is $P(=p_{2n-1}, \ldots, p_{n-1}, \ldots, p_0)$.

4.2 QCA Array Multipliers

4.2.1 Structural Design

Array multipliers that are well suited to QCA are studied and analyzed in this section. An array multiplier is formed by a regular two dimensional lattice of identical functional units so that the structure conforms to QCA technology without extra wire delay.

4.2.2 Schematic Design

The type I 4-bit by 4-bit array multiplier shown in Figure 4.2 has four columns and five rows in a somewhat irregular lattice. On the other hand, the type II 4-bit by 4-bit array multiplier shown in Figure 4.3 is formed by a regular lattice of identical functional units with four rows. The type I 4-bit by 4-bit array

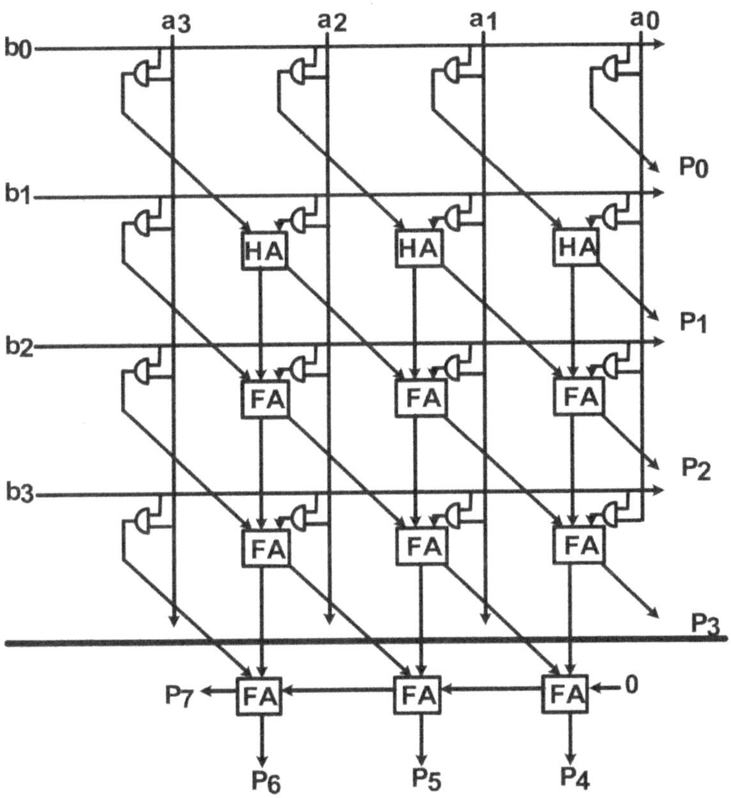

Figure 4.2 Schematic of a Type I array multiplier. (© 2010 IEEE. *From* [8].)

multiplier consists of 16 AND gates, three half adders (HAs), and nine full adders (FAs). The carry outputs from each FA go to the next row. The type II 4-bit by 4-bit array multiplier has 16 AND gates and 16 FAs. The carry outputs in this structure go to the left in the same row. One operand propagates from left to right in the type I array multiplier while the input signals in the type II multiplier go to the left. The type I array multipliers have a latency of $2N-2$ adder delays, while the type II array multipliers have a latency of $2N$ adder delays. Either a half adder or a FA is assumed to have one adder delay from any input to any output. Table 4.1 shows the required hardware for the two types of array multipliers.

4.2.3 Implementation of Array Multipliers with QCAs

The two different types of 4-bit by 4-bit array multipliers are implemented in Figures 4.4 and 4.5. They are extended to make larger (8-bit by 8-bit) multipliers as shown in Figures 4.6 and 4.7. The area of the type I array multiplier gets 4.35 times larger as the operand size is doubled due to its irregular lattice

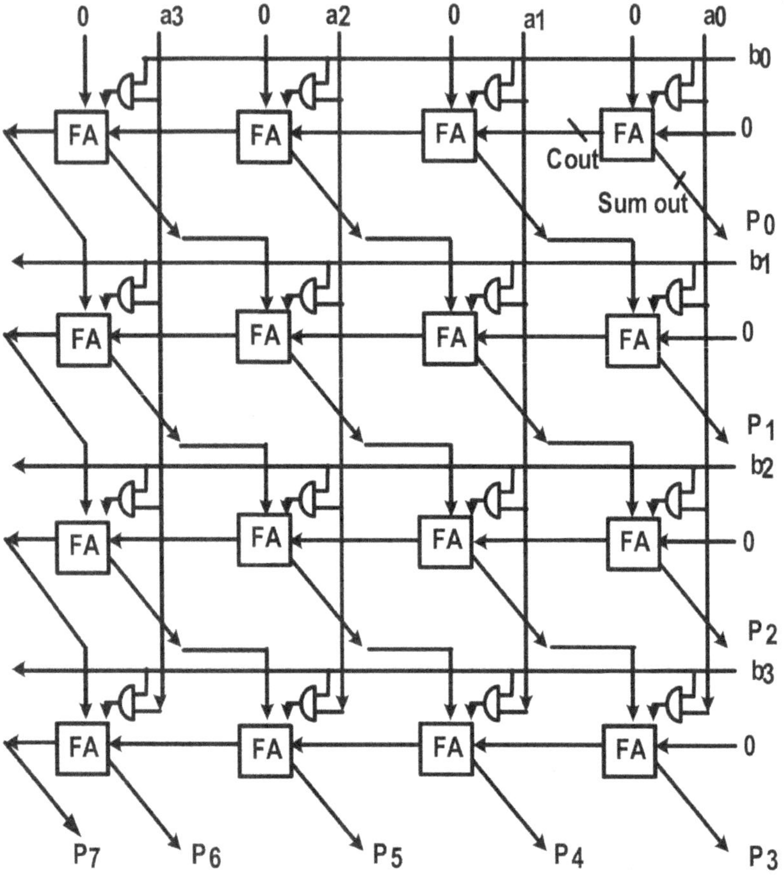

Figure 4.3 Schematic of a type II array multiplier. (© 2010 IEEE. *From* [8].)

Table 4.1
Required Hardware for the Array Multipliers

Multipliers	Number of FAs	Number of HAs	AND Gates
4-bit × 4-bit type I array	9	3	16
4-bit × 4-bit type II array	16	0	16
8-bit × 8-bit type I array	49	7	64
8-bit × 8-bit type II array	64	0	64
N-bit × N-bit type I array	$(N-1)^2$	$(N-1)$	N^2
N-bit × N-bit type II array	N^2	0	N^2

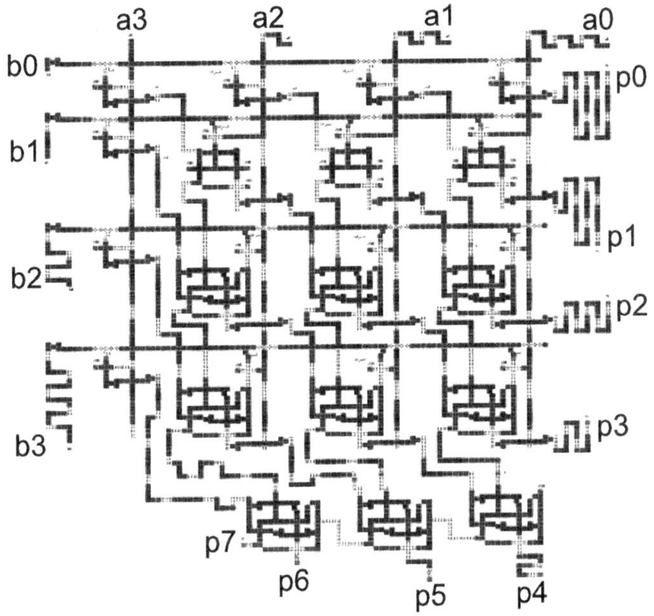

Figure 4.4 Layout of a 4-bit by 4-bit type I array multiplier. (© 2010 IEEE. *From* [8].)

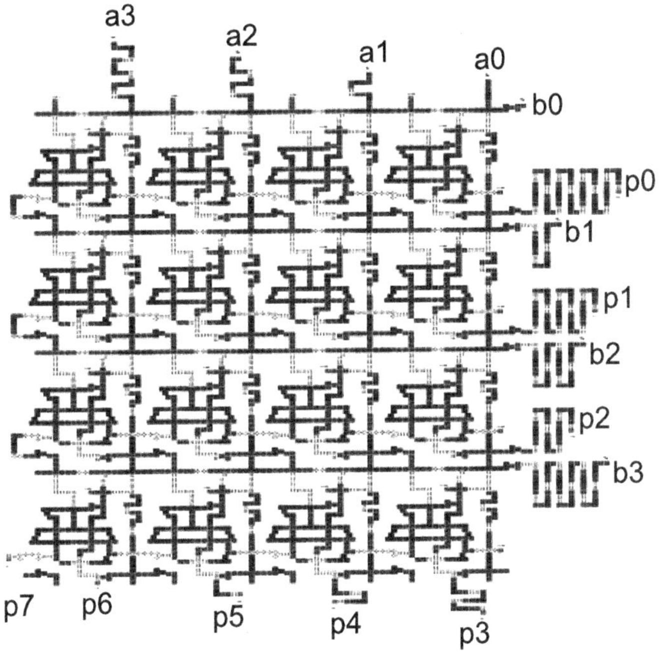

Figure 4.5 Layout of a 4-bit by 4-bit type II array multiplier. (© 2010 IEEE. *From* [8].)

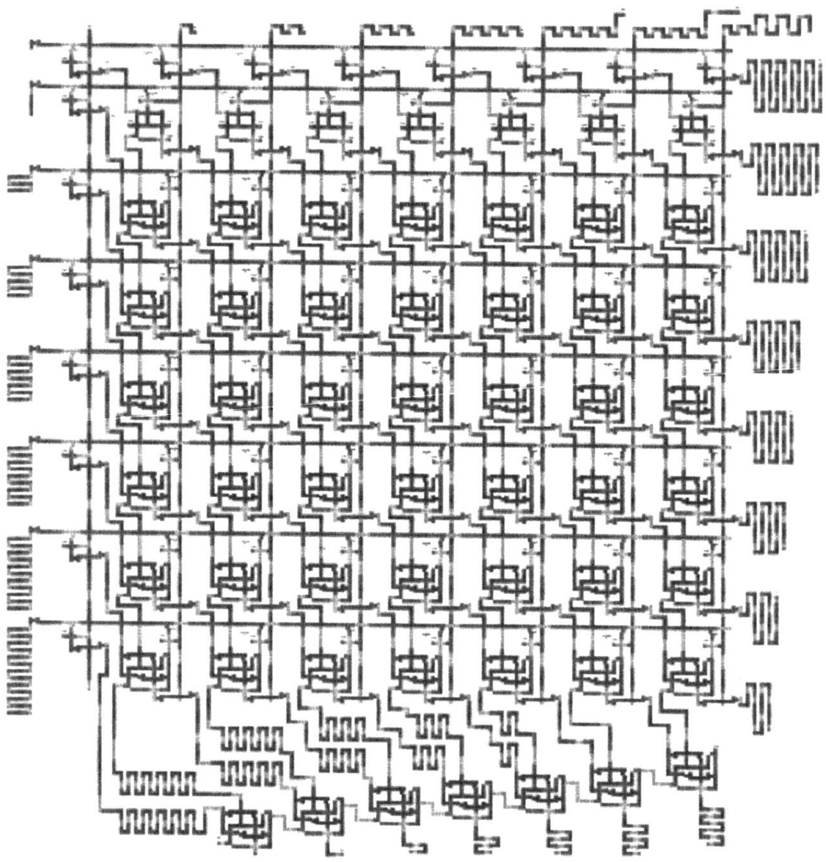

Figure 4.6 Layout of an 8-bit by 8-bit type I array multiplier. (© 2010 IEEE. *From* [8].)

and one more last row. The type I array multiplier has fewer cells than the type II array multiplier due to its fewer FAs. On the other hand, the type II array multiplier is less than four times larger as the operand size is doubled because of its identical unit cells.

4.3 Wallace and Dadda Multipliers For QCA

4.3.1 Introduction

To handle the repetitive addition, the Wallace [4] and Dadda [5] parallel tree multipliers use different strategies [6]. Wallace's strategy is to combine partial product bits at the earliest opportunity, while Dadda's method does the combining as late as possible, consistent with keeping the same critical path length through the carry save adder tree. Dadda's method yields a slightly simpler re-

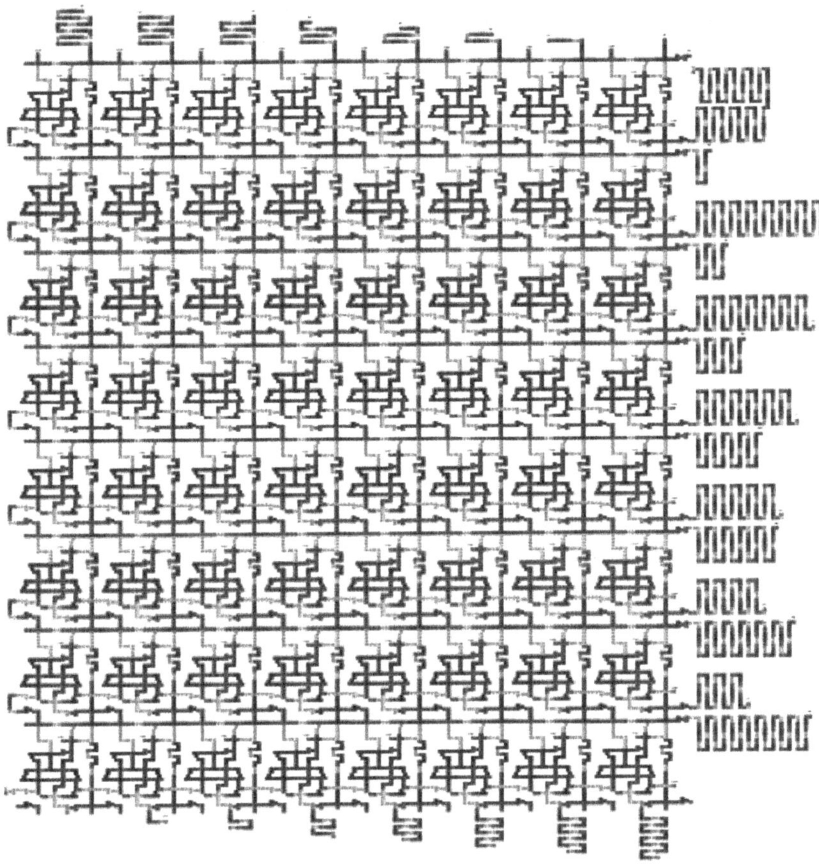

Figure 4.7 Layout of an 8-bit by 8-bit type II array multiplier. (© 2010 IEEE. *From* [8].)

duction tree and a slightly wider carry propagate adder. Figure 4.8 shows dot diagrams of 4-bit by 4-bit Wallace and Dadda multipliers that have two reduction stages. A dot indicates a partial product of the multiplication. Plain and crossed diagonal lines indicate the outputs of FAs and half adders, respectively. The block diagrams of 4-bit by 4-bit Wallace and Dadda multipliers are shown in Figures 4.9 and 4.10, respectively.

4.3.2 Schematic Design

Figure 4.11 shows the formation of the partial products for multiplication. The set of these partial products forms a bit matrix that is added in the second step. Each multiple is produced by a two-input AND gate. The AND gate is realized with a three-input majority gate that has one input tied to logic 0.

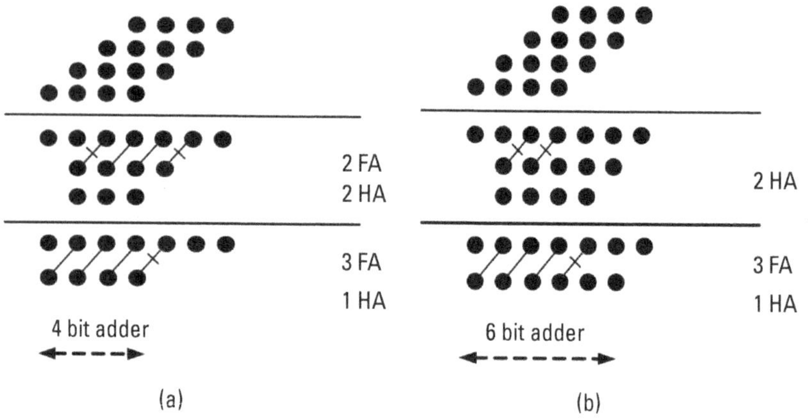

Figure 4.8 Dot diagrams of 4-bit by 4-bit reductions: (a) Wallace, (b) Dadda. (© 2009 IEEE. *From* [1].)

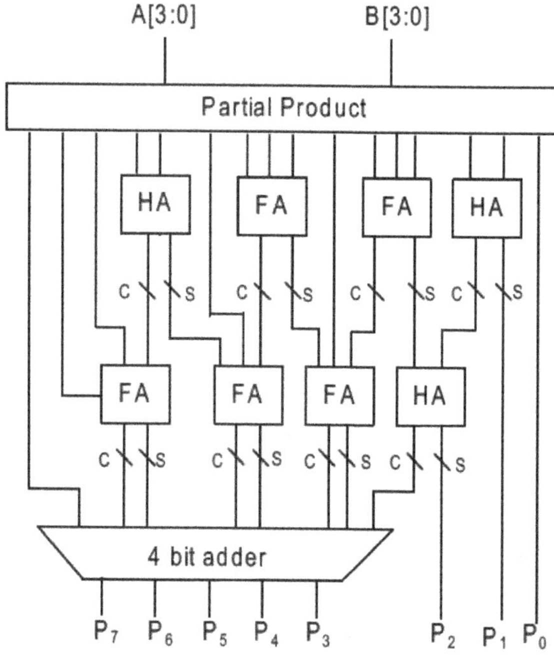

Figure 4.9 Block diagram of a 4-bit by 4-bit Wallace multiplier. (© 2009 IEEE. *From* [1].)

In each stage of the reduction, the Wallace multiplier conducts a preliminary grouping of rows into sets of three. Within each three-row set, FAs and half adders are used to reduce the three rows to two. Rows that are not part of a three row set are transferred to the next stage without modification. The bits of these rows are considered in the later stages. An 8-bit by 8-bit Wallace

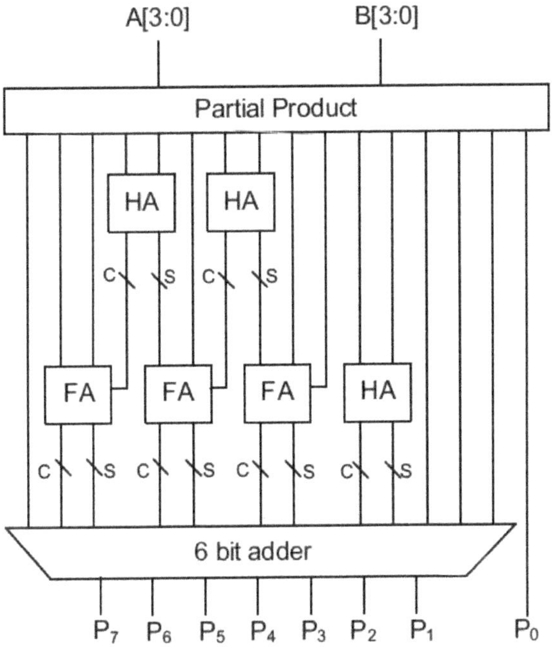

Figure 4.10 Block diagram of a 4-bit by 4-bit Dadda multiplier. (© 2009 IEEE. *From* [1].)

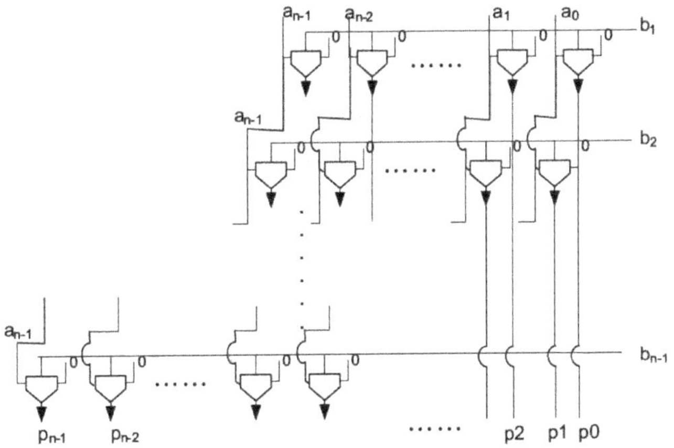

Figure 4.11 Generation of the partial products. (© 2009 IEEE. *From* [1].)

multiplier has four reduction stages and intermediate matrix heights of 6, 4, 3, and 2. A dot diagram of an 8-bit by 8-bit Wallace multiplier is shown in Figure 4.12.

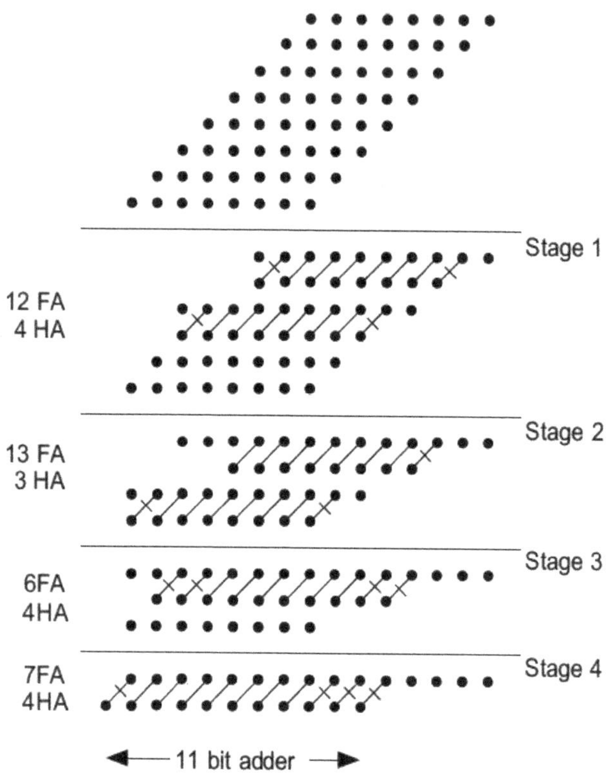

Figure 4.12 Dot diagram of an 8-bit by 8-bit Wallace multiplier. (© 2010 IEEE. *From* [8].)

The Dadda multiplier combines partial product bits as late as possible and has a sequence of intermediate matrix heights that minimizes the number of full and half adders needed. It is determined by working back from the last stage (i.e., the two-row matrix). Dadda's method is to apply FA and half adders to obtain a reduced matrix with no column heights greater than a specified limit. The Dadda limit sequence is 2, 3, 4, 6, 9, 13, 19, 28, 42, and so on. A dot diagram of an 8-bit by 8-bit Dadda multiplier with intermediate matrix heights of 6, 4, 3, and 2 is shown in Figure 4.13.

Table 4.2 shows the required hardware. Wallace multipliers have more FAs and half adders than Dadda multipliers in the reduction process, but have smaller final carry propagation adders.

4.3.3 Implementation with QCAs

The layouts of 4-bit by 4-bit Wallace and Dadda multipliers are shown in Figures 4.14 and 4.15, respectively. In these multipliers stair-like RCAs are used for synchronization and to make each stage pipelined. 3295 and 3384 cells are used

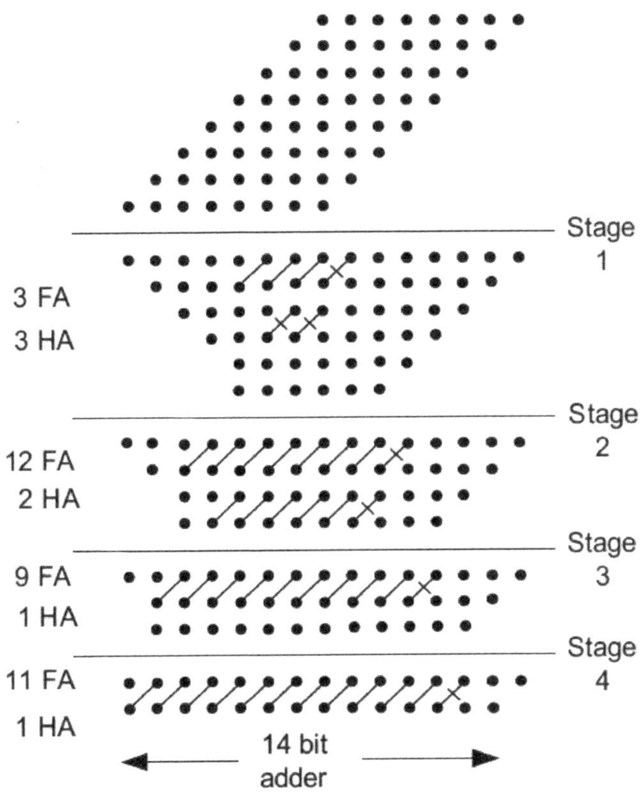

Figure 4.13 Dot diagram of an 8-bit by 8-bit Dadda multiplier. (© 2010 IEEE. *From* [8].)

Table 4.2
Required Hardware for the Multipliers

Multipliers	Number of FAs	Number of HAs	Final Adder Size
4 × 4 Wallace	5	3	4-bit
4 × 4 Dadda	3	3	6-bit
8 × 8 Wallace	38	15	11-bit
8 × 8 Dadda	35	7	14-bit

to make 4-bit by 4-bit Wallace and Dadda multipliers with areas of 7.39 μm^2 and 7.51 μm^2, respectively. 8-bit by 8-bit Wallace and Dadda multipliers with expandable partial products are shown in Figures 4.16 and 4.17. The 8-bit by 8-bit multipliers have about 10 times as many cells as 4-bit by 4-bit multipliers.

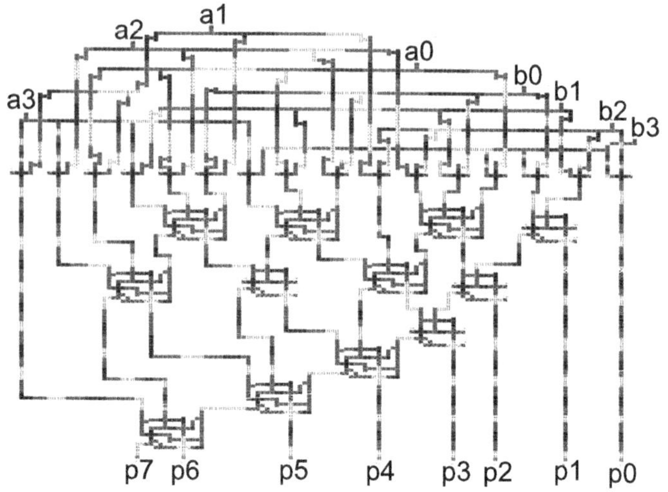

Figure 4.14 Layout of a 4-bit by 4-bit Wallace multipliers. (© 2009 IEEE. *From* [1].)

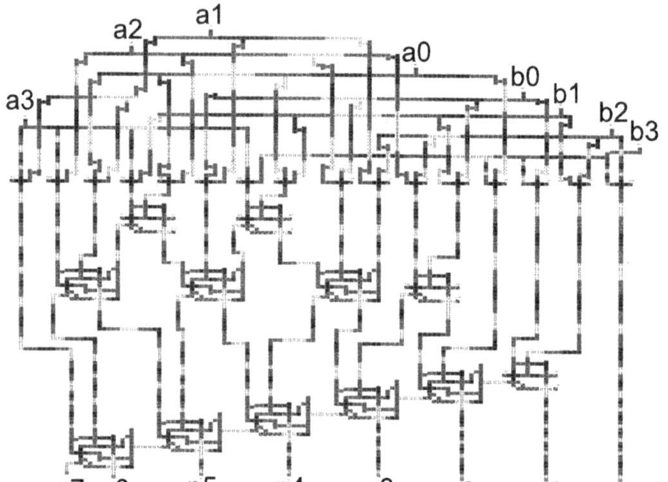

Figure 4.15 Layout of a 4-bit by 4-bit Dadda multiplier. (© 2009 IEEE. *From* [1].)

4.4 Quasi-Modular Multipliers For QCA

4.4.1 Quasi-Modular Multiplier Method

This section describes the design of quasi-modular parallel multipliers using the 4-bit by 4-bit Wallace and Dadda multipliers of the previous section as building blocks. Consider the multiplication of two n-bit magnitudes using modules that perform the multiplication of a k-bit magnitude by an l-bit magnitude. To

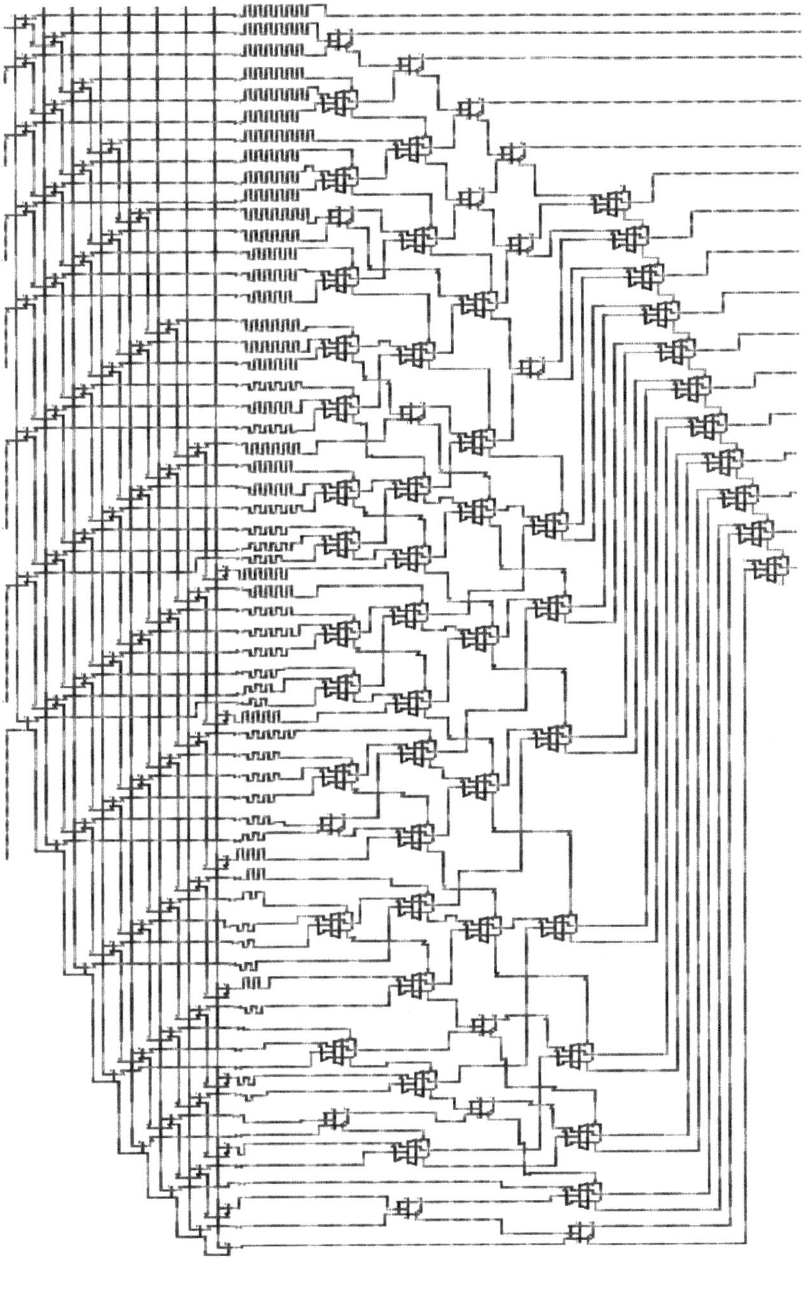

Figure 4.16 Layout of an 8-bit by 8-bit Wallace multiplier. (© 2010 IEEE. *From* [8].)

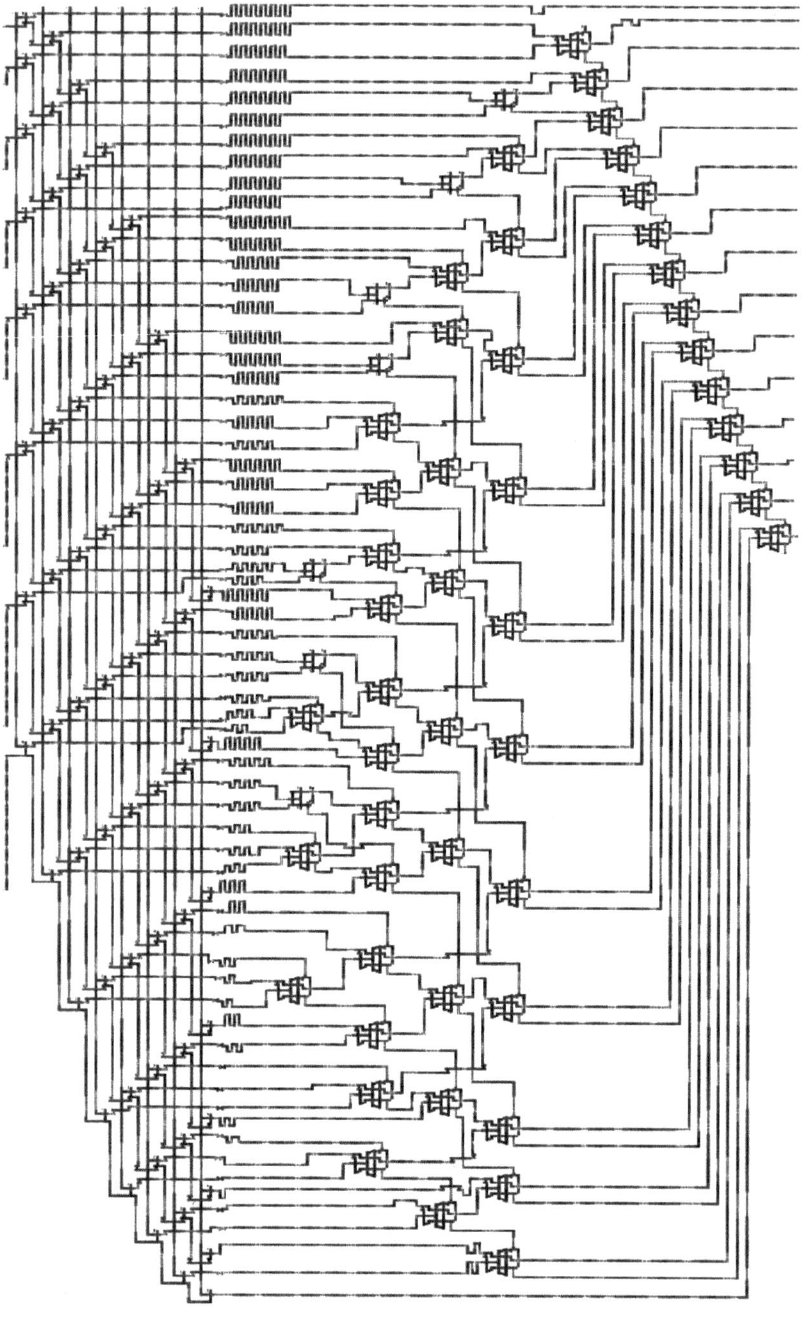

Figure 4.17 Layout of an 8-bit by 8-bit Dadda multiplier. (© 2010 IEEE. *From* [8].)

perform the $n \times n$ multiplication using these $k \times l$ modules, the operands are decomposed into radix 2k and 2l digits, respectively. Then the multiplication is:

$$\begin{aligned} P &= X \times Y \\ &= \sum x^{(i)} y^{(j)} 2^{ki+lj} \\ &= \sum p^{(i,j)} 2^{ki+lj} \end{aligned} \tag{4.2}$$

Thus, $(n/k) \times (n/l)$ modules are needed. The output of these modules, when suitably aligned, produces a quasi-modular product (QMP) bit-matrix.

4.4.2 Structural Design

For example, an 8-bit by 8-bit multiplier constructed using four (4-bit by 4-bit) modules is shown in Figure 4.18. The decomposition of the operands and the multiplication are:

$$x = x^{(1)} 2^4 + x^{(0)} \tag{4.3}$$

$$y = y^{(1)} 2^4 + y^{(0)} \tag{4.4}$$

$$x \times y = x^{(1)} y^{(1)} 2^8 + x^{(1)} y^{(0)} 2^4 + x^{(0)} y^{(1)} 2^4 + x^{(0)} y^{(0)} \tag{4.5}$$

This shows that large multipliers can be constructed with small modules. To make an 8-bit by 8-bit multiplier, four (4-bit by 4-bit) multiplier modules, eight FAs, and an 11-bit adder are used. In general an N-bit by N-bit multiplier can be made with four ($N/2$-bit by $N/2$-bit) multiplier modules, N FAs, and a $(3N/2 - 1)$-bit adder.

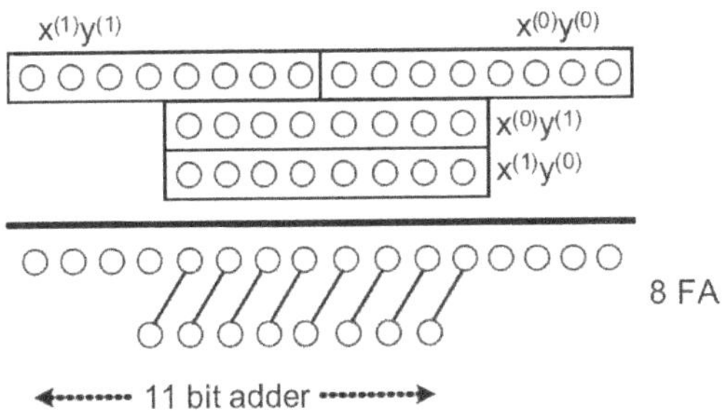

Figure 4.18 An 8-bit by 8-bit quasi-modular multiplier made using four (4-bit by 4-bit) modules, eight FAs and an 11-bit adder. (© 2009 IEEE. *From* [1].)

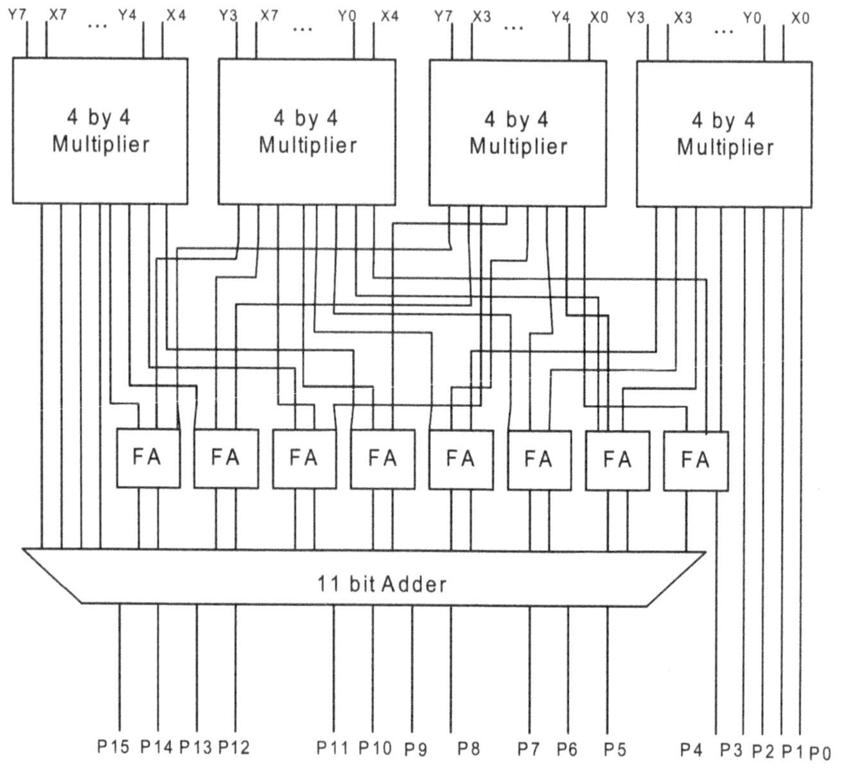

Figure 4.19 An 8-bit by 8-bit quasi-modular multiplier block diagram.

An 8-bit by 8-bit quasi-modular multiplier block diagram is shown on Figure 4.19. It is used for the implementations in this section.

4.4.3 Implementation with QCAs

Figures 4.20 and 4.21 show the layouts of the quasi-modular 8-bit by 8-bit multipliers that are based on the Wallace and Dadda 4-bit by 4-bit multipliers, respectively. The quasi-modular 8-bit by 8-bit Wallace and Dadda multipliers have 6.4% and 12.7% less area, respectively than the corresponding 8-bit by 8-bit Wallace and Dadda multipliers.

4.4.4 Simulation Results

Simulations were done with QCADesigner [7] assuming coplanar wire "crossings" and a maximum of 15 cells per clock zone. In the simulations, most default parameters for a bistable approximation as listed in Section 2.4.1.3 are used. The number of samples is determined to be 51,200. Figures 4.22 and 4.23 show simulation results of 4-bit by 4-bit multipliers.

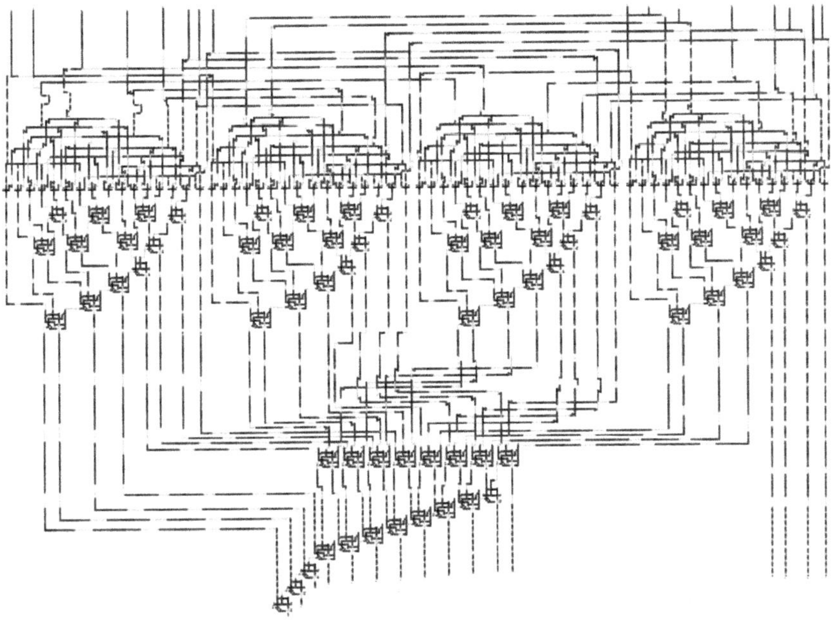

Figure 4.20 Layout of an 8-bit by 8-bit Wallace quasi-modular multiplier. (© 2009 IEEE. *From* [1].)

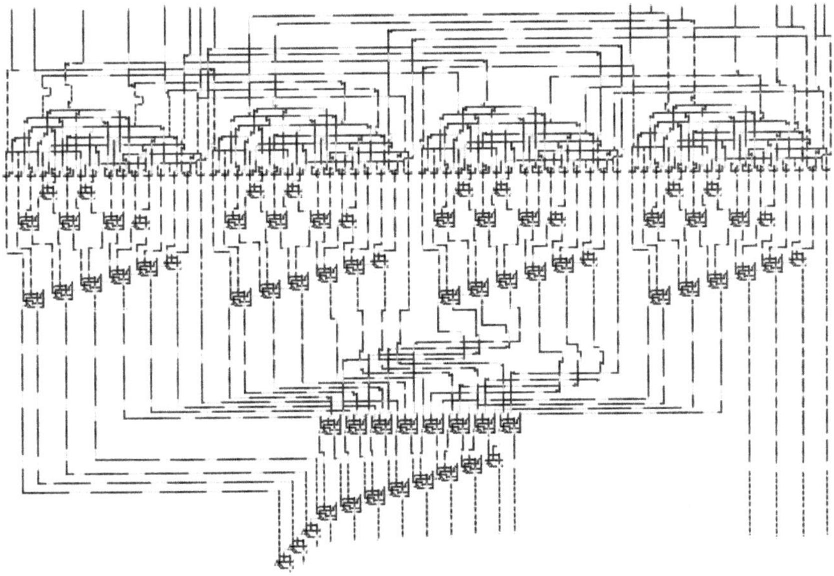

Figure 4.21 Layout of an 8-bit by 8-bit Dadda quasi-modular multiplier. (© 2009 IEEE. *From* [1].)

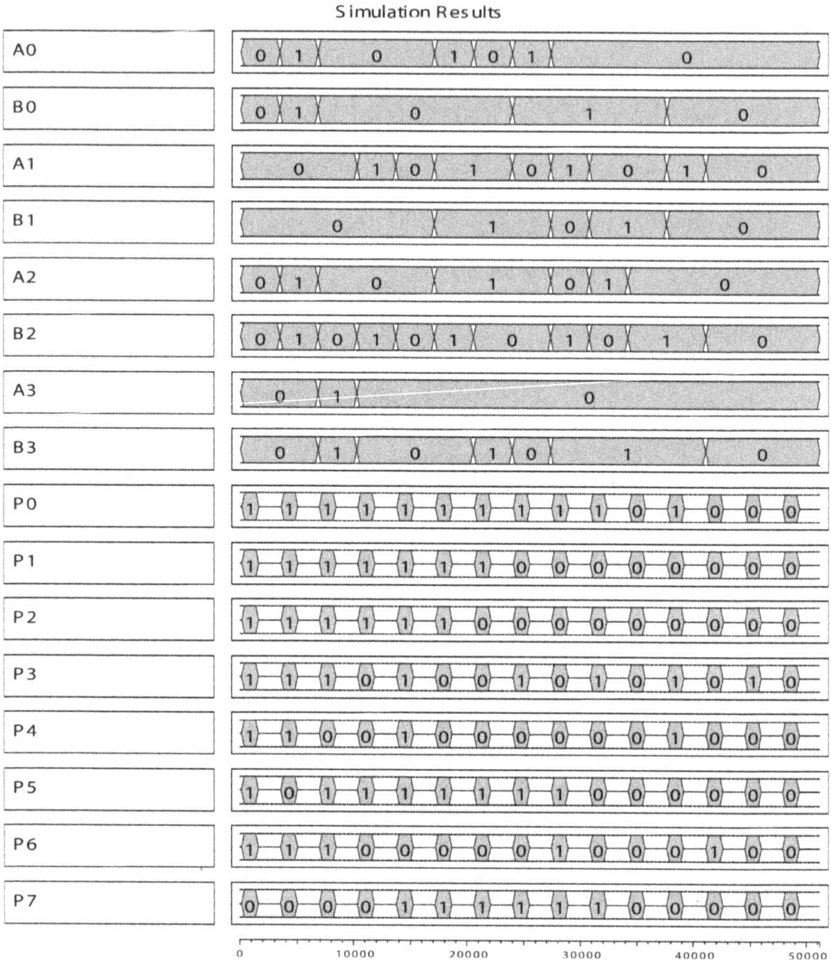

Figure 4.22 Simulation results for 4-bit by 4-bit Wallace multiplier.

Based on the number of components, it is expected that fast and small multipliers can be achieved with the quasi-modular multipliers as shown in Table 4.3. However, at present, these multipliers have a large wiring overhead that produces unused areas in the layout.

The 4-bit by 4-bit Wallace and Dadda multipliers have clock latencies of 10 and 12 cycles and require 3,295 cells and 3,384 cells, respectively with area that is comparable to the array multiplier. The 8-bit by 8-bit Wallace multiplier has more than four times the latency and 11 times the area of the 4-bit by 4-bit multiplier. The 8-bit by 8-bit Dadda multiplier has four times the latency and 12 times the area of the 4-bit by 4-bit Dadda multiplier. The quasi-modular 8-bit multipliers have more than three times the latency, about eight times

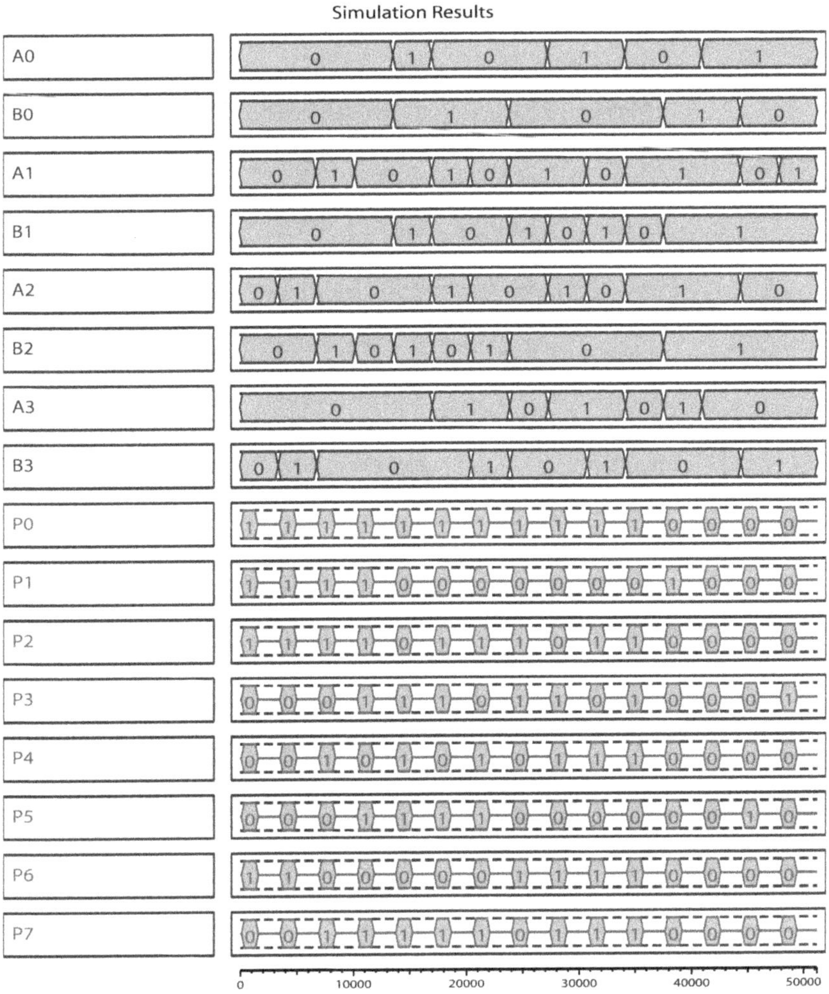

Figure 4.23 Simulation results for 4-bit by 4-bit Dadda multiplier.

as many cells and more than 10 times the area of the 4-bit multipliers. Even though the quasi-modular multipliers are much slower than what was expected as shown in Table 4.3, the 8-bit by 8-bit tree multipliers have large areas, many cells, and a longer delay than 8-bit by 8-bit quasi-modular multipliers. At least part of the explanation for this is that many of the wires are quite long as is evident in Figures 4.20 and 4.21. The long wires significantly affect the timing; 33.8% of the latency is due to the wiring. In the 8-bit by 8-bit multiplier cases, 47.7% and 44.6% of the total latency in the Wallace and Dadda multipliers are spent in partial products, respectively. Quasi-modular multipliers could be much faster if the wiring and layout problems were solved. These results

Table 4.3
Quasi-Modular Multipliers

Multipliers	Number of Modules	Number of HAs	Final Adder Size	Delay
8 × 8	4 (4 × 4)	8	11-bit	22
16 × 16	4 (8 × 8)	8	23-bit	46
32 × 32	4 (16 × 16)	8	47-bit	94
64 × 64	4 (32 × 32)	8	95-bit	190

Table 4.4
Comparison of QCA Multipliers

Multipliers	Latency	Area	Number of Cells
4 × 4 Array I	14	5.15 μm^2	2,956
4 × 4 Array II	14	6.02 μm^2	3,738
4 × 4 Wallace	10	7.39 μm^2	3,295
4 × 4 Dadda	12	7.51 μm^2	2,284
8 × 8 Array I	30	22.57 μm^2	13,839
8 × 8 Array II	30	21.49 μm^2	15,106
8 × 8 Wallace	44	87.47 μm^2	33,894
8 × 8 Dadda	47	92.69 μm^2	34,903
8 × 8 quasi-modular (Wallace)	36	82.18 μm^2	26,499
8 × 8 quasi-modular (Dadda)	38	82.19 μm^2	26,973

show that systolic and simple structures are needed. In addition, if they can be realized, multilayer wire crossings might mitigate the wire burden.

4.5 Comparison of QCA Multipliers

Table 4.4 shows a comparison of the simulation results for 4-bit by 4-bit and 8-bit by 8-bit array, Wallace, Dadda and quasi-modular multipliers. The various 4-bit by 4-bit multipliers have a clock latency of 10 to 14 clock cycles and require 2,956 to 3,738 cells. The 8-bit by 8-bit multipliers have roughly four times the latency and nine times the cell count of the 4-bit by 4-bit multipliers. The 8-bit by 8-bit quasi-modular multipliers have less latency and fewer cells than the 8-bit by 8-bit Wallace and Dadda multipliers. All of the 8-bit by 8-bit multipliers are much slower and larger than would be expected from CMOS multipliers. The 4-bit array multipliers are slower and smaller than the Wallace and Dadda multipliers. The 8-bit by 8-bit array multipliers have twice the latency and roughly four times the area and cell count of the 4-bit by 4-bit array multipliers. These results show that the most significant factor in the performance is the wiring. Also the number of cells is proportional to the latency.

Thus it seems that array multipliers are the best choice for QCA implementation. The latency is least (for all but the smallest multipliers), and the area is much less than Wallace, Dadda, and quasi-modular multipliers.

4.6 Conclusion

This chapter describes fully parallel multipliers with large operand sizes in addition to quasi-modular multipliers that mitigate the need for complex irregular tree structure architectures. Parallel multiplier trees including array multipliers, Wallace and Dadda multipliers have been constructed and analyzed in QCA technology. To facilitate modular design and to accommodate large word sizes, quasi-modular parallel multipliers are proposed and compared with the previous designs. It is found that even though the quasi-modular multipliers are much slower than expected, 8-bit by 8-bit tree multipliers have large areas, many cells and a longer delay than 8-bit by 8-bit quasi-modular multipliers. It has been shown again that long wires significantly affect timing with 33.8% of the latency due to the wiring. Quasi-modular multipliers could be much faster if the wiring and layout problems were solved.

Unlike CMOS technology, structures that can reduce wiring overhead are desirable and have advantages in QCA technology. For example, array multipliers are faster and have less area than column compression multipliers due to their conformability with QCA technology, which avoids extra wire delay. The array multipliers are the best choice for QCA implementation, as the latency is least and the area is much less than that of the other three multipliers. To take advantage of super pipelined QCA building blocks, simple systolic structures need to be developed.

References

[1] Kim, S., and E. Swartzlander, Jr., "Parallel Multipliers for Quantum-Dot Cellular Automata," in *Proceedings of the IEEE Nanotechnology Materials and Devices Conference*, 2009, pp. 68–72.

[2] Pezaris, S. D., "A 40-ns 17-bit by 17-bit array multiplier," *IEEE Transactions on Computers*, Vol. C-20, 1971, pp. 442–447.

[3] Baugh, C. R., and B. A. Wooley, "A two's complement parallel array multiplication algorithm," *IEEE Transactions on Computers*, Vol. C-22, 1973, pp. 1045–1047.

[4] Wallace, C. S., "A suggestion for a fast multiplier," *IEEE Transactions on Electronic Computers*, Vol. EC-13, 1964, pp. 14–17.

[5] Dadda, L., "Some schems for parallel multipliers," *Alta Frequenza,* Vol. 34, 1965, pp. 349–356.

[6] Bickerstaff, K. C., M. J. Schulte, and E. E. Swartzlander Jr., "Parallel reduced area multipliers," *The Journal of VLSI Signal Processing,* Vol. 9, 1995, pp. 181–191.

[7] Walus, K., et al., "QCADesigner: A Rapid Design and Simulation Tool for Quantum-Dot Cellular Automata," *IEEE Transactions on Nanotechnology,* Vol. 3, 2004, pp. 26–31.

[8] Kim, S., and E. E. Swartzlander, Jr., "Multipliers with Coplanar Crossings for Quantum-Dot Cellular Automata," *10th IEEE Conference on Nanotechnology (IEEE-NANO),* 2010, pp. 953–957.

5

QCA Dividers

Inwook Kong, Seong-Wan Kim, and Earl E. Swartzlander, Jr.

5.1 Introduction

Division is the slowest and most complex of the common arithmetic operations due to its inherently serial nature. Nonetheless division algorithms are constantly researched, because division operations cannot be overlooked in many applications. This chapter presents the designs of two very different types of dividers, a digit recurrent divider and a convergent divider. Digit recurrent dividers (similar to paper and pencil division) are widely used in conventional technology, because their implementation is relatively simple. Convergent dividers, where an initial estimate of the quotient is refined iteratively, are used where high speed is required and when one or more multipliers are available. Section 5.2 presents the design of the restoring binary divider. Section 5.3 presents the design of a Goldschmidt convergent divider.

5.2 Digit Recurrent Divider[1]

5.2.1 Types of Digit Recurrent Dividers

Digit recurrent dividers were implemented in software in the earliest general purpose digital computers. The basic types are restoring, nonrestoring, and SRT (named after Sweeney, Robertson, and Tocher). For a QCA implementation, a restoring binary design was selected.

1. Section 5.2 is based on [1].

5.2.2 Conventional Restoring Binary Divider Architecture

Restoring division [2–4] is performed in a recursive procedure by the following formula:

$$P^{(i+1)} = r \times P^{(i)} - q_{i+1} \times D \tag{5.1}$$

where, $i = 0, 1,..., n - 1$ is the recursion index, $P^{(i)}$ is the partial remainder in the ith iteration, D is the divisor, and r is the radix. The initial partial remainder $P^{(0)}$ equals the dividend, and $P^{(n)}$ is the final remainder. Binary (r=2) restoring division operates on fixed-point fractional numbers and depends on the following assumptions: $0 \leq P^{(i+1)} < D$, $D < 1$, the allowable quotient digits q_{i+1} are from the digit set {0, 1}. The quotient digit is selected by performing a sequence of subtractions and shifts. Each time, D is subtracted from the partial remainder $r \times P^{(i)}$ until the difference becomes negative. Then D is added back to the negative difference, which is called restoring. For binary division, $r = 2$ and the quotient digit can be determined as follows:

$$q_{i+1} = \begin{cases} 0, \text{ if } 2P^{(i)} < D \\ 1 \text{ if } 2P^{(i)} \geq D \end{cases} \tag{5.2}$$

The partial remainder is obtained by one left shift of $P^{(i)}$ and by one subtraction: $P^{(i+1)} = 2P^{(i)} - D$. If it is positive, then $q_{i+1} = 1$ else $q_{i+1} = 0$ and a restoring addition is required. The addition is needed to restore the correct partial remainder, $P^{(i+1)} = P^{(i+1)} + D = 2P^{(i)}$. Thus, performing binary restoring division requires at least one subtraction and may require one addition to determine each quotient digit. A nonperforming restoring binary division needs only one subtraction per quotient bit, because it uses a register to restore the partial remainder. Figure 5.1 shows a conventional restoring divider. This conventional divider is a restoring division scheme, and hence it has to load the trial difference in the partial remainder register in all cases restoring the remainder to its correct value by a compensating addition step. As a result, this scheme is slow. Wire-dominant QCA characteristics will adversely affect the overall latency.

5.2.3 Restoring Binary Divider

The design and construction of an iterative QCA array for parallel division uses many replicated units for the comparison of the partial remainder and divisor to reduce QCA wires. The shift is realized by pipelined QCA wiring. To improve the throughput of the QCA array divider, inherent pipelining can be used by inserting latches on the output for each row of cells. In general, a k by k restoring array divider (RAD) receives a $2k$-bit dividend and k-bit divisor

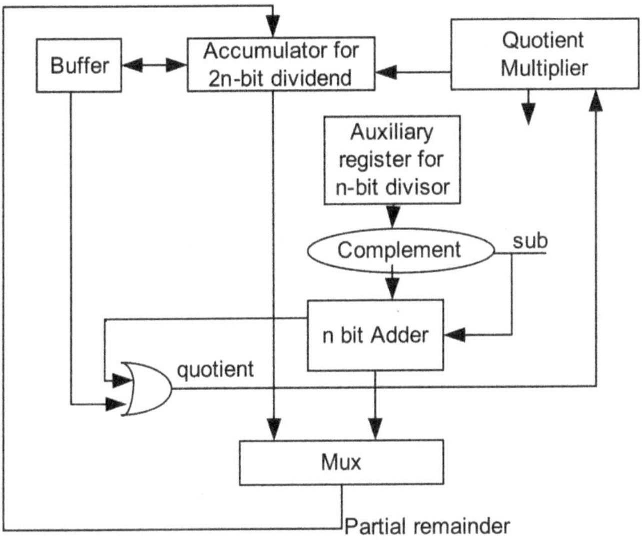

Figure 5.1 Restoring divider block diagram. (© 2011 IEEE. *From* [1].)

and produces a k-bit quotient and a $2k$-bit remainder (including k leading 0s). Consider the division of a $2k \times k$ RAD with $k = 6$:

- Dividend $z = .z_1z_2z_3z_4z_5z_6z_7z_8z_9z_{10}z_{11}z_{12}$;
- Divisor $d = .d_1d_2d_3d_4d_5d_6$;
- Quotient $q = .q_1q_2q_3q_4q_5q_6$;
- Remainder $s = .000000s_7s_8s_9s_{10}s_{11}s_{12}$.

Figure 5.2 shows an example of a binary restoring division. A block diagram of a 6-bit by 6-bit RAD with controlled subtractor cells is shown in Figure 5.3. Each cell consists of a full subtractor and a two-input MUX that is the basic element of the RAD. When the control input $P = 1$, the divider input d is subtracted from the partial remainder and the difference cell_{out} is passed down to find the difference between the previous partial remainder and the divisor. Otherwise, when the control input $P = 0$, the MUX in a row cell is triggered so the partial remainder input bits are passed down unchanged. The difference cell_{out} is a function of P:

$$\text{cell}_{\text{out}} = \begin{cases} z \oplus d \oplus c_{in}, & \text{if } P = 1 \\ z, & \text{if } P = 0 \end{cases} \tag{5.3}$$

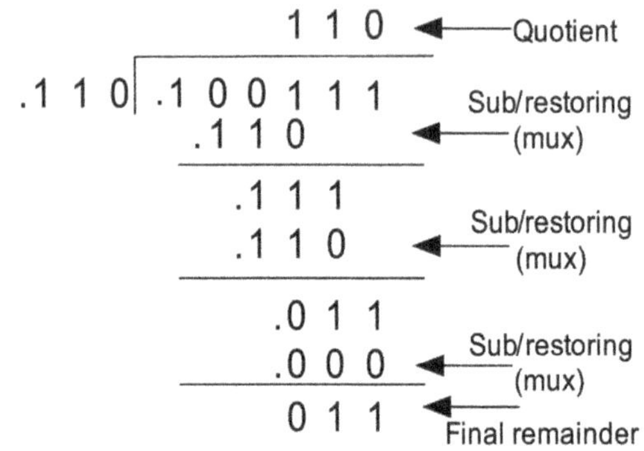

Figure 5.2 An example of a 3-bit by 3-bit binary restoring division. (© 2011 IEEE. *From* [1].)

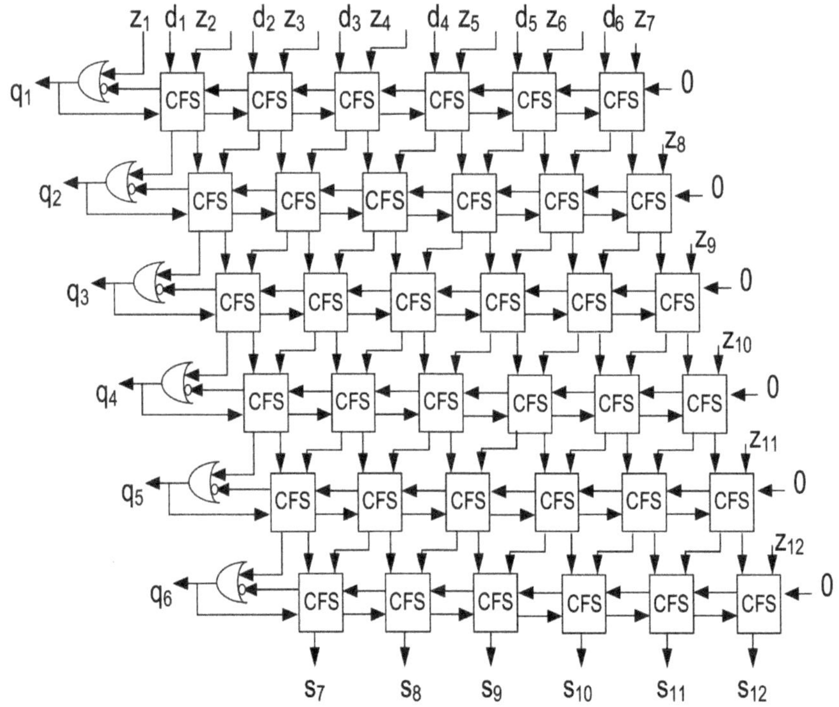

Figure 5.3 Block diagram of a 6-bit by 6-bit RAD. (© 2011 IEEE. *From* [1].)

Quotients q_i are obtained from the left of each row. $q_i = \overline{b}_{out} + z_i = P$ where z_i is the bit shifted output of the $(i - 1)$th iteration. If $z_i = 1$ or $b_{out} = 0$, then a subtraction is performed.

5.2.4 Implementation of the Restoring Divider

In order to design a robust circuit, the divider has been designed using coplanar wire crossings with the design guidelines suggested in [5–7]. The design guidelines in [5] are kept except for a limitation on majority gate outputs. Robust operation of majority gates is attained by limiting the maximum number of cells that are driven by the output, which is verified using the coherence vector method. The maximum number of cells for each circuit component in a clock zone is determined by simulations with sneak noise sources. For example, the maximum number of cells for a simple wire is 14, and the minimum is 2.

5.2.4.1 Basic Elements for RAD

The RAD is composed of controlled full subtractor cells that have one full subtractor and one two-input MUX. The subtractor design using majority gates and inverters can be expressed as:

$$D_i = A_iB_iC_i + \bar{A}_i\bar{B}_iC_i + A_i\bar{B}_i\bar{C}_i + \bar{A}_iB_i\bar{C}_i \tag{5.4}$$

$$D_i = M\left(M\left(\bar{A}_i, B_i, C_i\right), M\left(\bar{A}_i, \bar{B}_i, C_i\right), \bar{C}_i\right) \tag{5.5}$$

$$C_{i+1} = \bar{A}_iB_i + \bar{A}_iC_i + B_iC_i \tag{5.6}$$

$$C_{i+1} = M\left(\bar{A}_i, B_i, C_i\right) \tag{5.7}$$

A full subtractor can be designed using only three majority gates and three inverters. Figure 5.4 shows the full subtractor schematic and layout. The full subtractor takes 3 inputs, (*A*, *B*, and *C*) and gives two outputs, (the difference, D_i, and the borrow output, C_{i+1}. Both outputs (i.e., the difference and borrow signals) have a latency of one clock cycle.

The two input MUX shown in Figure 5.5 is another cell element required for the controlled full subtractor cell. When control signal, *S*, is 0, input *A* is output and when control signal, *S*, is 1, input *B* is output with a latency of one clock cycle.

5.2.4.2 Controlled Full Subtractor Cell

The controlled full subtractor (CFS) cell shown in Figure 5.6 basically performs the operation $z - d$ if $P_{\text{in}} = 1$ to find the difference between the previous partial remainder and divisor. The borrow signal is output after a latency of 2 clock cycles. Instead of shifting the partial remainder left to form $2P^{(i)}$, the remainder

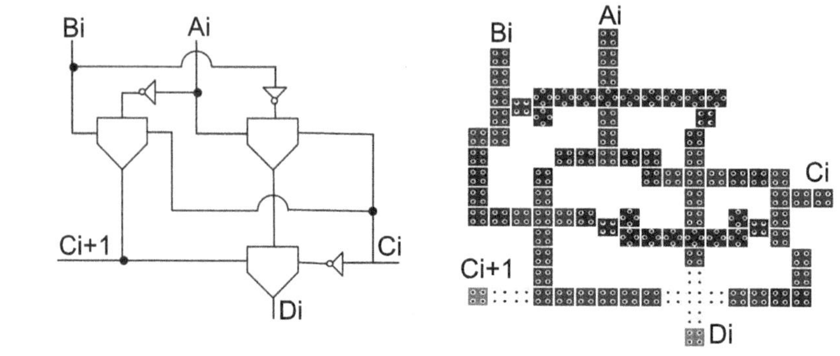

(a) (b)

Figure 5.4 A full subtractor: (a) schematic, (b) layout. (© 2011 IEEE. *After* [1].)

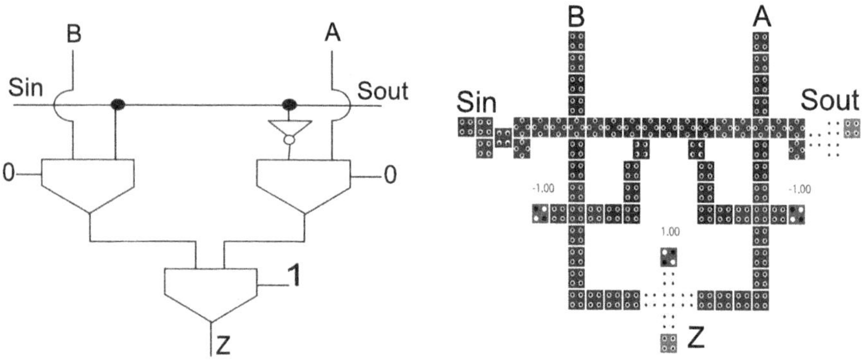

(a) (b)

Figure 5.5 A 2:1 MUX: (a) schematic, (b) layout. (© 2011 IEEE. *After* [1].)

is fixed and the divisor is shifted right along the divisor delay lines; quotient bits are obtained from the left of each row. The outputs $cell_{out}$ and borrow have latencies of three clock cycles and two clock cycles, respectively. To make an array divider, this cell unit is modified for synchronization.

5.2.4.3 Timing Analysis

The RAD resembles an array multiplier, but it has different delay characteristics due to borrow signal propagation along the row and due to the quotient signal that is fed back to each cell. From the rightmost column, CFS cells have 22, 18, 14, 10, 6, and 2 latches between the full subtractor and the MUX to implement the pipelining. Each cell has 21 delay latches for the divisor to pass down to the next row. A timing block diagram for a 6-bit by 6-bit RAD is shown on Figure 5.7.

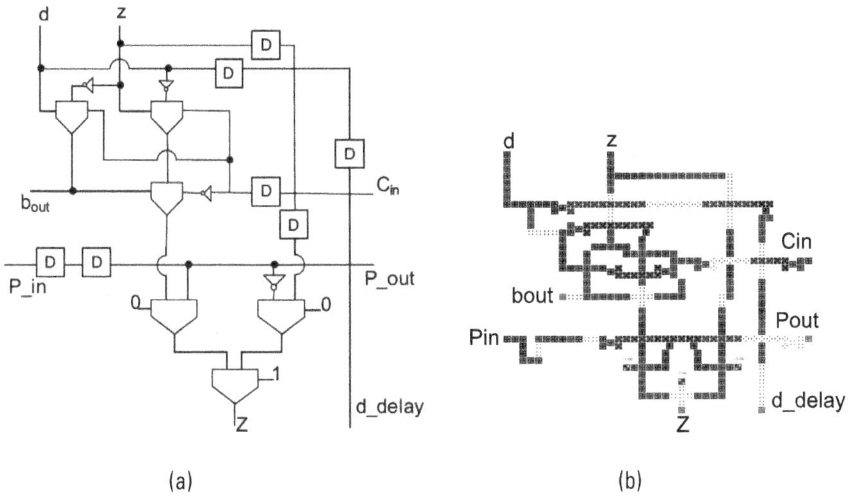

Figure 5.6 CFS cell: (a) schematic and (b) layout. (© 2011 IEEE. *From* [1].)

5.2.5 Simulation Results

5.2.5.1 Restoring Divider Results

Simulations were done with QCADesigner [8] assuming coplanar "crossings." In the simulations, most default parameters for a bistable approximation as listed in Section 2.4.1.3 are used. The number of samples is determined to be 25,600 for the 3-bit by 3-bit RAD and 156,000 for the 6-bit by 6-bit RAD.

Each cell with different delays is tested exhaustively, and the full integration is verified with consecutive input vectors:

- Dividend {2672, 3335, 2228, 2346, 3670, 2328};
- Divisor {51, 56, 52, 50, 62, 44}.

Correct data is output with a latency of 145 clock cycles. The values of the Quotient are {52, 59, 42, 46, 59, 52}, and the Remainder is {20, 31, 44, 46, 12, 40}. As the size is increased, RAD cells become larger along the column lines. The full simulation took 42 hours for the 12-bit by 12-bit RAD.

Layouts of the 6-bit by 6-bit and 12-bit by 12-bit RADs are shown in Figures 5.8 and 5.9, respectively.

5.2.5.2 Analysis of QCA Digit Recurrent Restoring Dividers

Table 5.1 shows a comparison of three sizes of RADs. The dividers get much larger as the operand size increases. The latency of the RAD is $4N^2 + 1$ so it increases by a factor of about 4 as the operand size is doubled. Even though the

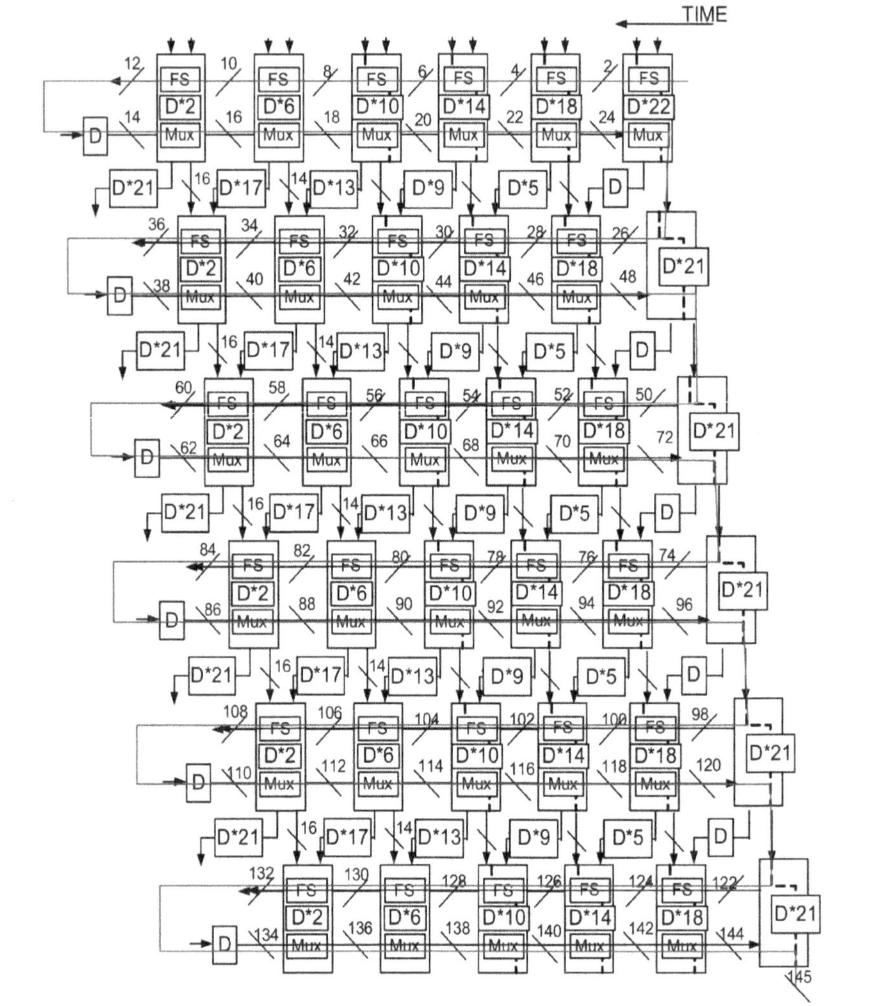

Figure 5.7 Timing block diagram for a 6-bit by 6-bit RAD. (© 2011 IEEE. *From* [1].)

latency is large, a new division can be started on each clock so the throughput is extremely high. The area and the cell count increase at an even faster rate than the latency.

5.3 Convergent Divider[2]

Large iterative computational circuits such as convergent dividers are difficult to build using QCAs with conventional sequential circuit design methods that

2. Section 5.3 is based on [9].

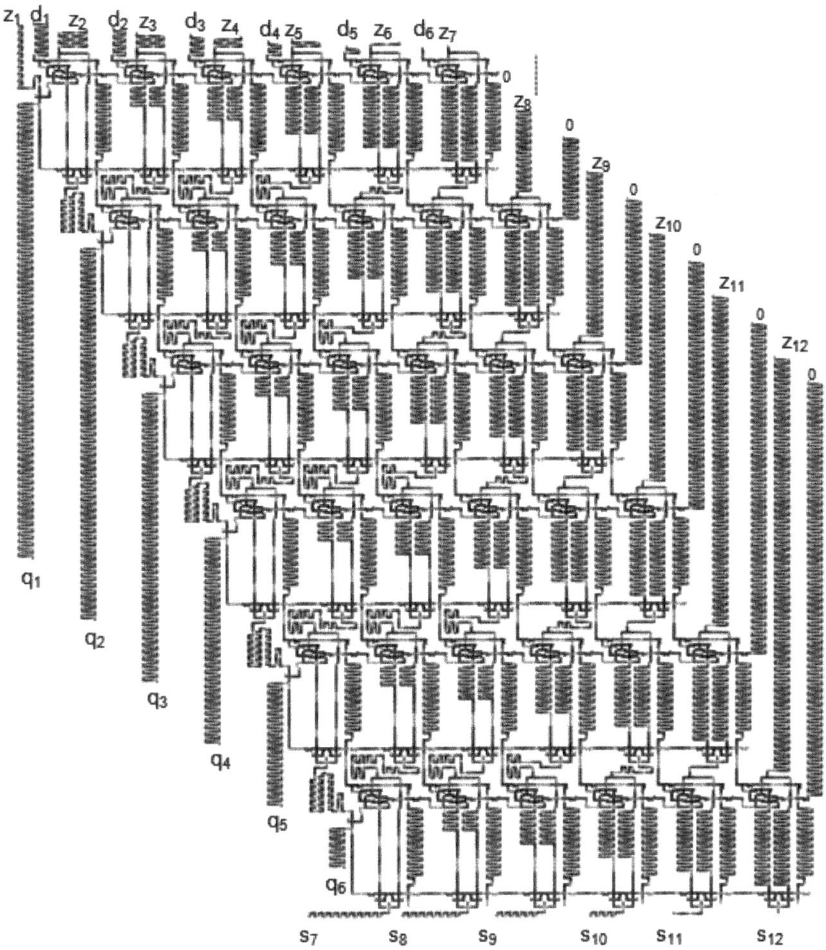

Figure 5.8 Layout of the 6-bit by 6-bit RAD. (© 2011 IEEE. *From* [1].)

are based on state machines. For this application, due to QCA wire delays, state machines have problems due to long delays between the state machine and the units to be controlled. Even a simple 4-bit microprocessor that has been implemented with QCA [10] was done without using a state machine. Due to the difficulty of designing sequential circuits, there has been little research into using QCAs to realize large iterative computational units, such as dividers.

In this section, a design is presented for a convergent divider using the Goldschmidt algorithm implemented with a data tag architecture to solve the difficulty in designing iterative computation units. Section 5.3.1 describes the Goldschmidt iterative division algorithm. Section 5.3.2 presents the data tag method. Section 5.3.3 details an implementation of the Goldschmidt divider using the proposed method. Finally, Section 5.3.4 presents simulation results.

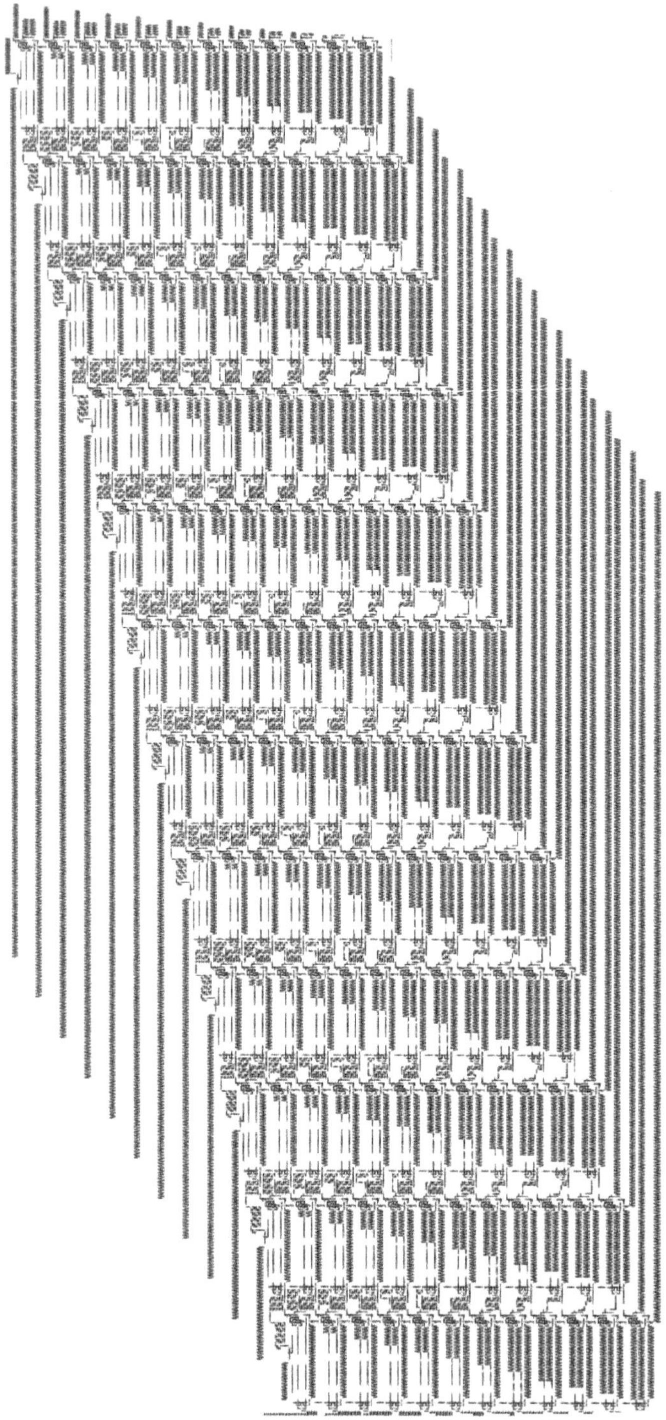

Figure 5.9 Layout of the 12-bit by 12-bit RAD. (© 2011 IEEE. *From* [1].)

Table 5.1
Comparison of QCA Dividers

Divider Size	Latency	Area	Cell Count
3×3	37	15 μm^2	6,451
6×6	145	86 μm^2	42,236
12×12	577	740 μm^2	301,395

5.3.1 The Goldschmidt Division Algorithm

In Goldschmidt division [11, 12], an approximate quotient converges toward the true quotient by multiple iterations. While an early software version was used for decimal division in the Harvard Mark IV computer [13], the first hardware implementation was for the IBM System/360 Model 91 [14].

The division operation can be viewed as the manipulation of a fraction. The numerator (N) and the denominator (D) are each multiplied by a sequence of numbers so that the value of D approaches 1 and the value of N approaches the quotient. In the first step, both N and D are multiplied by F_0, an approximation to the reciprocal of D. Often F_0 is produced by a reciprocal table with very limited precision, thus, the product of D times F_0 is not exactly 1, but has an error, ε.

Therefore, the first approximation of the quotient is:

$$Q = \frac{N \times F_0}{D \times F_0} = \frac{N_0}{D_0} = \frac{N_0}{1 - \varepsilon} \tag{5.8}$$

At the next iteration, N_0 and D_0 are multiplied by F_1, which is given by:

$$F_1 = (2 - D_0) = 2 - (1 - \varepsilon) = 1 + \varepsilon \tag{5.9}$$

Note that often the one's complement of D_i is used for F_{i+1} as that avoids the need to do a subtraction. The result is a slight (1 LSB) increase in the error, which very slightly reduces the rate of convergence.

$$Q = \frac{N_0 \times F_1}{D_0 \times F_1} = \frac{N_0(1+\varepsilon)}{(1-\varepsilon)(1+\varepsilon)} = \frac{N_0(1+\varepsilon)}{1-\varepsilon^2} = \frac{N_1}{D_1} \tag{5.10}$$

At the ith iteration, F_i is as follows:

$$F_i = (2 - D_{i-1}) = 1 + \varepsilon^{2^{i-1}}, \text{ for } i > 0 \tag{5.11}$$

As the iterations continue, N_i converges toward Q with quadratic precision, which means that the number of correct digits doubles on each iteration.

To illustrate the Goldschmidt division, consider the following example: $Q = 0.6/0.75$. From a lookup table, the approximate reciprocal of D (i.e., 0.75) is $F_0 = 1.3$:

$$Q = \frac{N \times F_0}{D \times F_0} = \frac{0.6 \times 1.3}{0.75 \times 1.3} = \frac{0.78}{0.975} = \frac{N_0}{D_0}$$

Then $F_1 = 2 - D_0 = 2 - 0.975 = 1.025$:

$$Q = \frac{N_0 \times F_1}{D_0 \times F_1} = \frac{0.78 \times 1.025}{0.975 \times 1.025} = \frac{0.7995}{0.999375} = \frac{N_1}{D_1}$$

Then $F_2 = 2 - D_1 = 2 - 0.999375 = 1.000625$:

$$Q = \frac{N_1 \times F_2}{D_1 \times F_2} = \frac{0.7995 \times 1.000625}{0.999375 \times 1.000625} = \frac{0.799999685}{0.999999609395} = \frac{N_2}{D_2}$$

The errors between N_0, N_1, N_2, and Q are 0.02, 0.0005, and 0.0000003125, respectively. This shows that the value of N_i converges quadratically to the value of Q.

A block diagram of a Goldschmidt divider for realization with CMOS technology is shown in Figure 5.10. It uses MUXs and latches that are controlled by a state machine to implement the iterations. Although this architecture is suitable for CMOS, distances from the state machine to the various elements are of varying lengths. Since QCA "wires" are implemented by strings of cells, delays vary according to the length of the wires. Due to the irregular wire delays, it is difficult to synchronize the inputs to the elements. This presents difficulties in QCA. Even in CMOS implementations, as the feature size continues to shrink, control signals that travel over varying distances are expected to become a problem.

5.3.2 The Data Tag Method for Iterative Computation

5.3.2.1 The Data Tag Method

To resolve the problem of state machines, a data tag method is used as shown in Figure 5.11. In this method, data tags travel with the data and local tag decoders (TD in the figure) generate control signals for the computational circuits (i.e., COMP1, COMP2, COMP3, etc.). The tags travel with the data through the

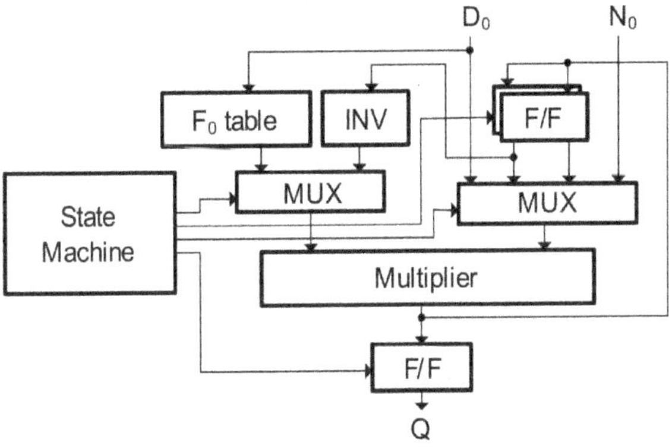

Figure 5.10 Goldschmidt divider block diagram for CMOS. (© 2009 IEEE. *After* [9].)

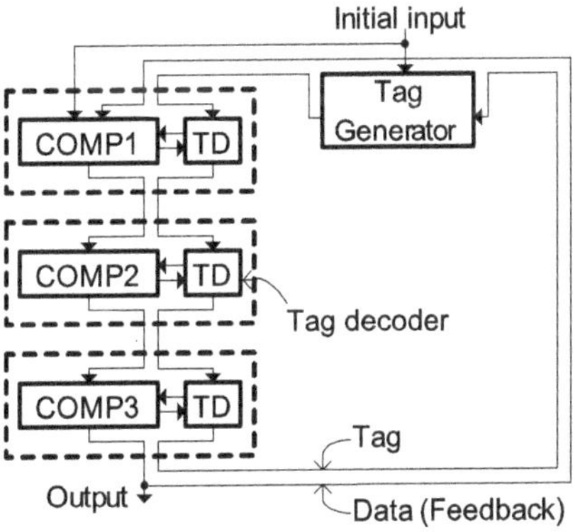

Figure 5.11 Computation unit implementation using data tags (TD = tag decoder). (© 2009 IEEE. *After* [9].)

same number of pipeline stages as the corresponding computational circuits, and local tag decoders generate control signals appropriate to each datum. Since the tags travel with the data and local tag decoders produce the control signals for the units, the synchronization issues that are a problem in state machines are significantly mitigated. In QCA, the data tag method can be implemented very efficiently since the delays to keep the data tags synchronized to the data are generated inherently via the QCA gates and wires.

The data tag method is also appropriate for CMOS as it allows freedom in laying out the various functional units since the signal transit time from the state machine to the functional units is no longer an issue. With CMOS, the computational circuits are generally implemented with far fewer pipeline stages than what is used for QCA, which means that fewer pipeline stages are needed to keep the data tags synchronized with the data.

Another advantage of the data tag architecture is that each datum on a data path can be processed differently according to the tag information. For example, in typical Goldschmidt dividers controlled by a state machine, a new division cannot be started until the previous division is completed. There are many pipeline stages in QCA computational circuits, and most stages may be idle during iterations. With the data tag method, each piece of data on a data path can be processed by the operation that is required at that stage. Since divisions at different stages are processed in a time-skewed manner, new divisions can be started while previous divisions are in progress as long as the initial pipeline stage of the data path is free. As a result, the throughput can be increased to a level that is much greater than that which is implied by the latency.

5.3.2.2 Goldschmidt Dividers Using the Data Tag Method

Two Goldschmidt dividers (one for 12-bit data and one for 24-bit data) have been designed using the data tag method. The flow chart is shown in Figure 5.12. Both dividers use the same low-precision ROM that gives 4-bit values for F_0. The 12-bit divider performs three iterations, the first with a value from the ROM followed by two iterations with the one's complement of D_i for F_{i+1}. The 24-bit divider performs four iterations, the first with a value from the ROM followed by three iterations with the one's complement of D_i for F_{i+1}.

The flow chart realizes the steps from Section 5.3.1. To start a new division, the tag generator issues a new tag (DT = 1) for the data. On the first

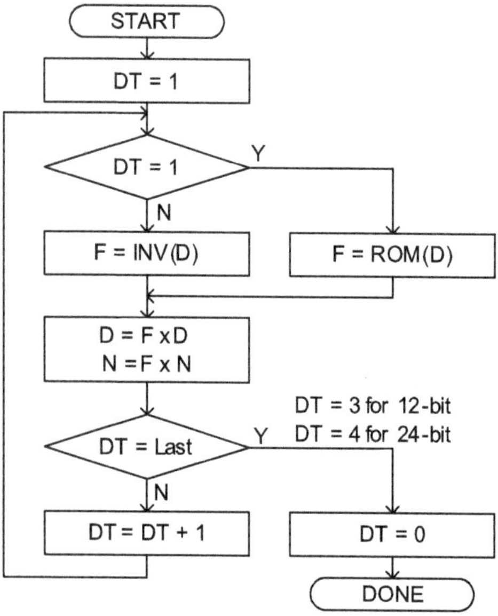

Figure 5.12 Flow charts of the 12-bit and 24-bit Goldschmidt dividers using data tags.

iteration, the factor is obtained from the ROM and used to multiply the denominator and numerator. For the second and third iterations, the factor is obtained by inverting the current value of the denominator. After the third or fourth iteration, the value of the numerator is output as the quotient. The local tag decoders control the MUXs according to the tag associated with the data. Once a division has started, it progresses through the required iterations, irrespective of any other divisions that are being performed.

5.3.3 Implementation of the Goldschmidt Divider

The Goldschmidt dividers have been designed using coplanar wire crossings with the design guidelines described in Section 5.2.4. A block diagram of both dividers is shown on Figure 5.13. The main elements are the ROM and the multiplier. In addition there are the tag generator, tag decoders, a word-wide inverter, and a few MUXs. The differences between the 12-bit and the 24-bit dividers are the width of the data paths, the inverters, the latches and the multiplier. *D* and *N* are input sequentially into the divider on successive clocks. The *Cmd* signal is asserted together with *D*, and a new tag is generated from the tag generator.

Then *N* is entered. The tag decoders control the MUXs and the latches using the tag that is associated with *D*. During the first iteration (used to normalize the denominator to a value that is close to 1), the MUXs are set so that *D* and *N* are multiplied sequentially by F_0 from the reciprocal ROM. After the first denominator normalization step is completed, the tag is incremented by

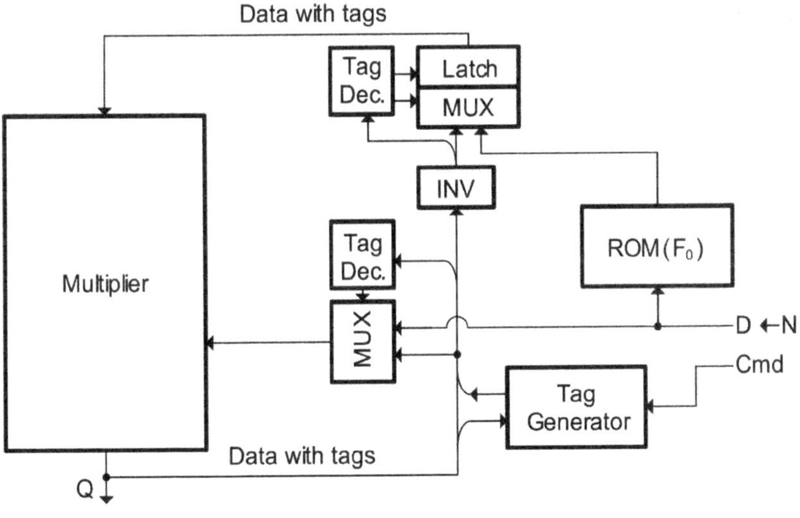

Figure 5.13 Block diagram of the Goldschmidt divider using the data tag method. (© 2009 IEEE. *After* [9].)

the tag generator. During the subsequent iterations, the MUXs select D or N from the outputs of the multiplier and F_i, which is computed by inverting the bits of D. (As noted in Section 5.3.1, this one's complement operation approximates 2 – D with an error of one LSB.) After three or four iterations, the final values of D and N have been computed, so N (the approximate value of Q) is output, and the tag generator eliminates the tag.

5.3.3.1 Tag Generators

The tag generators create the original tags and increment the tags on each iteration. The tag generator for the 12-bit multiplier creates a 2-bit tag, while the tag generator for the 24-bit multiplier creates a 3-bit tag. The tag generators for the 12-bit and 24-bit dividers are shown in Figures 5.14 and 5.15, respectively. Both generators have been implemented efficiently using majority logic reduction [15, 16]. A new tag ($TAG[1:0]$ =01 or $TAG[2:0]$ =001) is generated when the *Cmd* signal is asserted. In order to differentiate between D_i and N_i, the data tag is associated with only D_i, the first datum of a D_i, N_i pair. N_i may be associated with a dummy data tag since QCA wires for the data tags are not reset during start-up. Therefore, the dummy data tag is eliminated by two AND gates as shown in Figure 5.14. If a tag arrives at the tag generator after an iteration step, the numerical value of the tag is incremented for the next iteration. After a division is completed, the data tag is eliminated.

5.3.3.2 Tag Decoder and MUXs for D_i and N_i

The tag decoder and the MUXs for D_i and N_i for the 12-bit divider are implemented as shown in Figure 5.16. When $TAG[1 : 0]$is 01, the MUXs select D, N from the input data port. Since the MUXs pass N one clock after

D, the MUX selection signal in the tag decoder is held for two clock cycles. Since the MUXs for the less significant data have to be enabled earlier than other MUXs for the pipelined operation of the multiplier, the tag decoder is located near the MUX for B[0], and the control signals for the MUXs are in the reverse direction of the data flow. The schematics of the tag decoder and the MUXs for D_i and N_i for the 24-bit divider are the same as shown in Figure 5.16 because the tag decoder output is asserted when $TAG[2:0]$ is X01.

5.3.3.3 MUXs for F_i

The MUXs for F_i select the correct value of F_i (from either the ROM or the word-wide inverters) and hold the value for two clock periods. Thus, they require latches in order to hold F_i for two cycles. During the two cycles, D_i and N_i are multiplied by the value of F_i that is held by the latches. The decoder for the 12-bit divider is shown in Figure 5.17. The latch is implemented as an SR latch using a majority gate [17]. They are triggered when $TAG[1: 0]$ is not 00. The decoder for the 24-bit divider is shown in Figure 5.18.

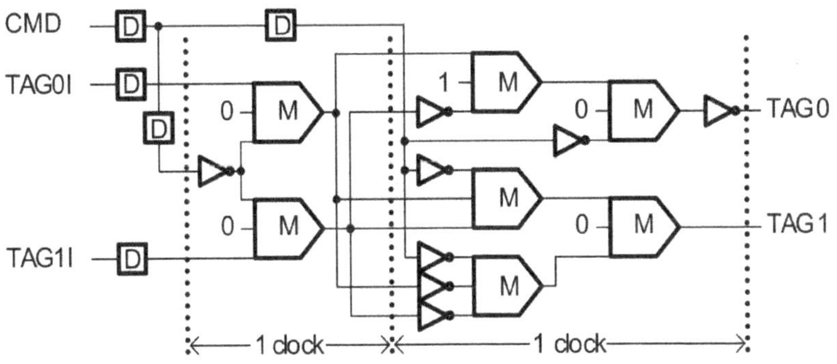

(a)

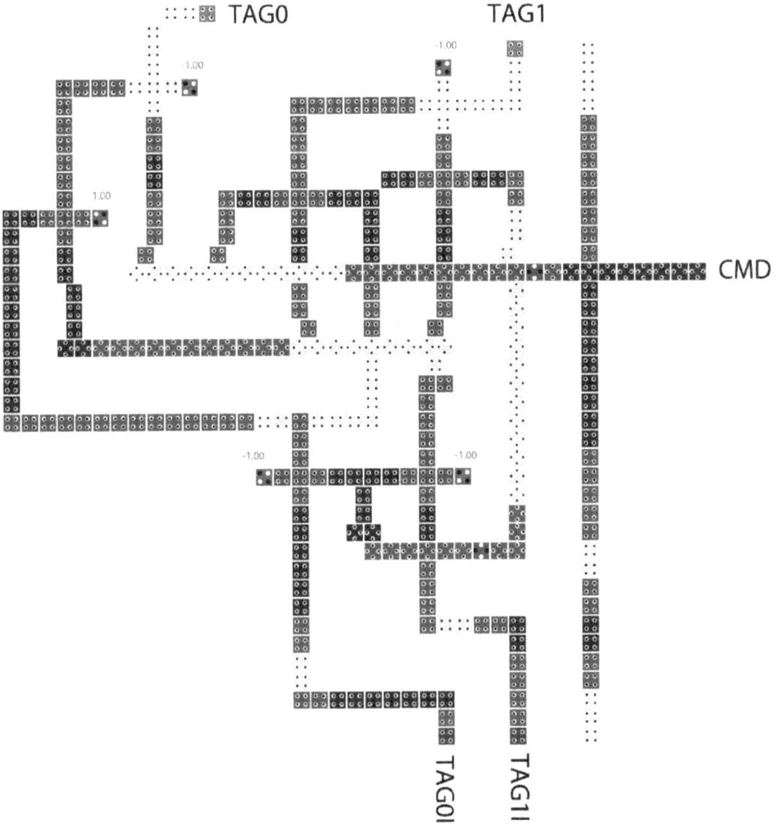

(b)

Figure 5.14 Tag generator for 12-bit divider: (a) schematic, (b) layout. (© 2009 IEEE. *After* [9].)

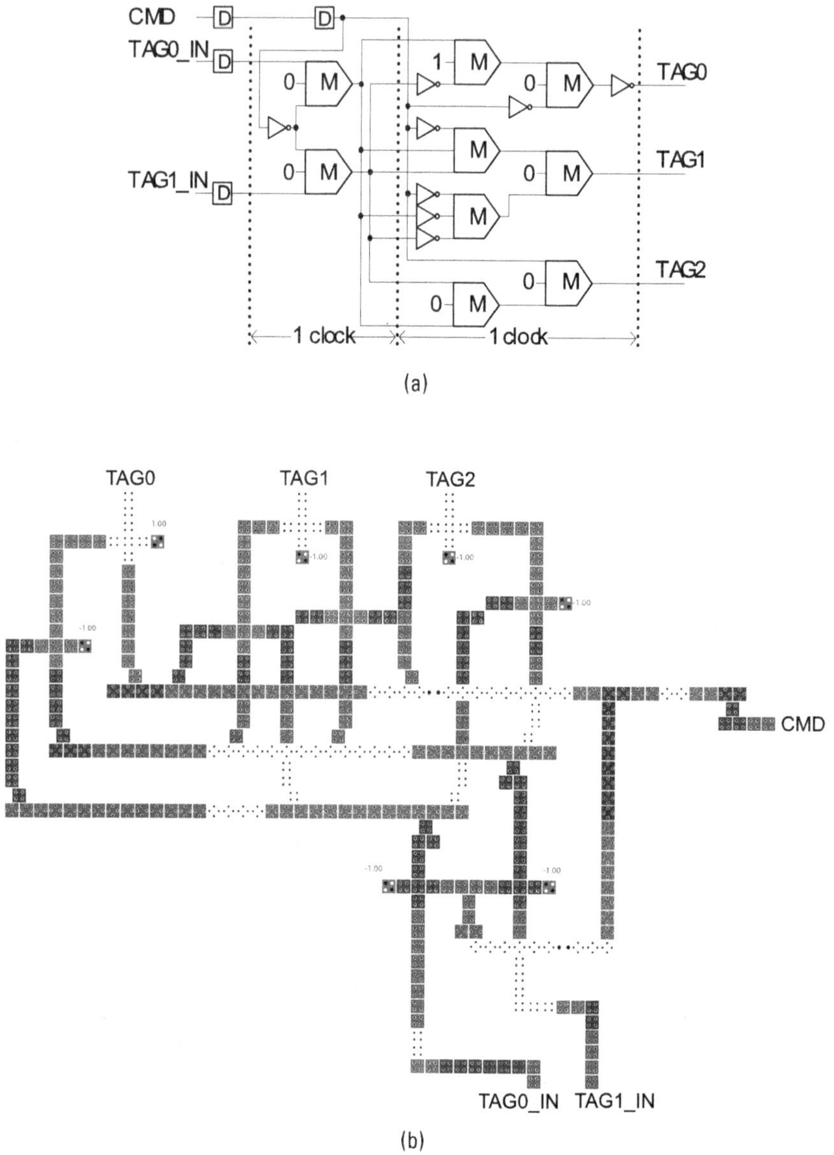

Figure 5.15 Tag generator for 24-bit divider: (a) schematic, (b) layout.

5.3.3.4 2^3- by 3-bit ROM Table

The 2^3- by 3-bit reciprocal ROM consists of a 3-bit decoder and an 8 by 3 ROM array as shown on Figure 5.19. All the ROM cells have the same access time, seven cycles. The data is programmed by setting one input of the OR gate inside each ROM cell. Both the 12-bit and the 24-bit dividers use the same ROM. Since the range of $D_i[0:11]$ is $0.5 \leq D < 1$ for the Goldschmidt divi-

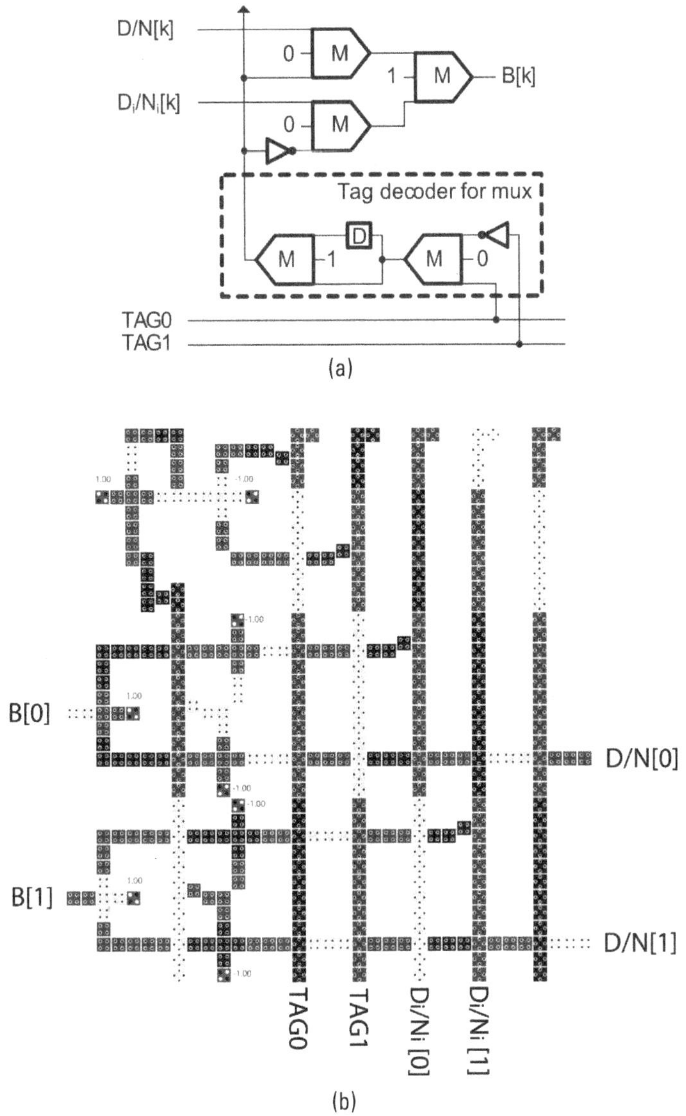

Figure 5.16 Tag decoder for MUXs for D_i, N_i: (a) schematic, (b) layout. (© 2009 IEEE. *After* [9].)

sion, $D_i[0:1]$ is always 01, so $D_i[2:4]$ is used as the input to the 3-bit ROM. Similarly, the ROM output is $F_0[1:3]$ since $F_0[0]$ is always 1. Thus, an 8 by 3 ROM implements a 32-word by 4-bit table.

5.3.3.5 Array Multiplier

The 12-bit divider uses a 12-bit by 12-bit array multiplier since array multipliers are attractive for QCA as shown in Chapter 4 and [18, 19]. The basic cell of

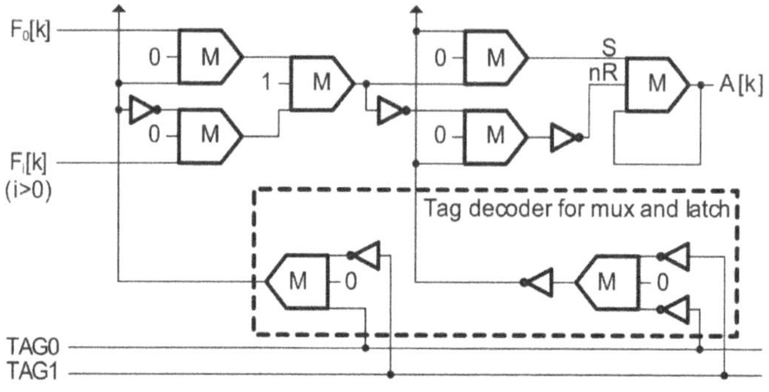

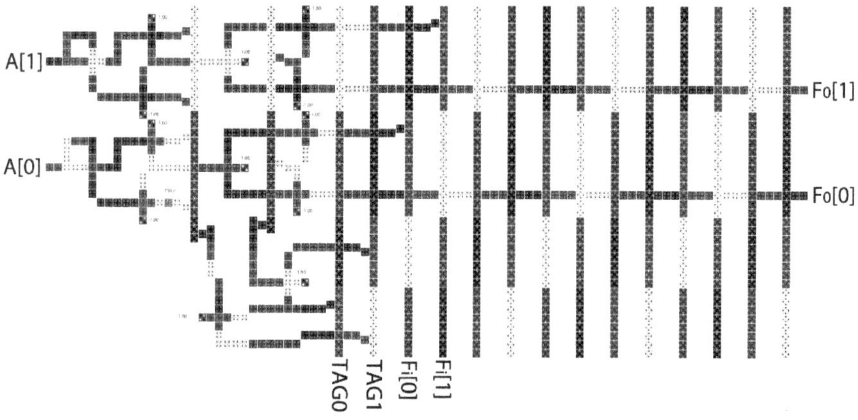

Figure 5.17 Tag decoder for F_i for the 12-bit Goldschmidt divider: (a) schematic, (b) layout for k=0, 1. (© 2009 IEEE. *After* [9].)

the multiplier shown in Figure 5.20 is a full adder implemented with three majority gates [20]. The cell has signal delays of 1 clock cycle for the carry output, two clock cycles for the sum, and an area of 20 by 29 cells. The schematic for the 4 × 4-bit multiplier using the multiplier cell is shown in Figure 5.21. The 12-bit by 12-bit multiplier has two inputs, A[11 :0] and B[11 :0], and a most significant output, M[11: 0]. The latency of an N-bit by N-bit array multiplier is $4N - 2$, so the 12-bit array multiplier has a latency of 46 clock cycles, which dominates the latency in an iteration. The unused least significant outputs are not left unconnected since that would violate the QCA design guidelines. Additional dummy cells are attached to the unused outputs for robust transfer of the signals.

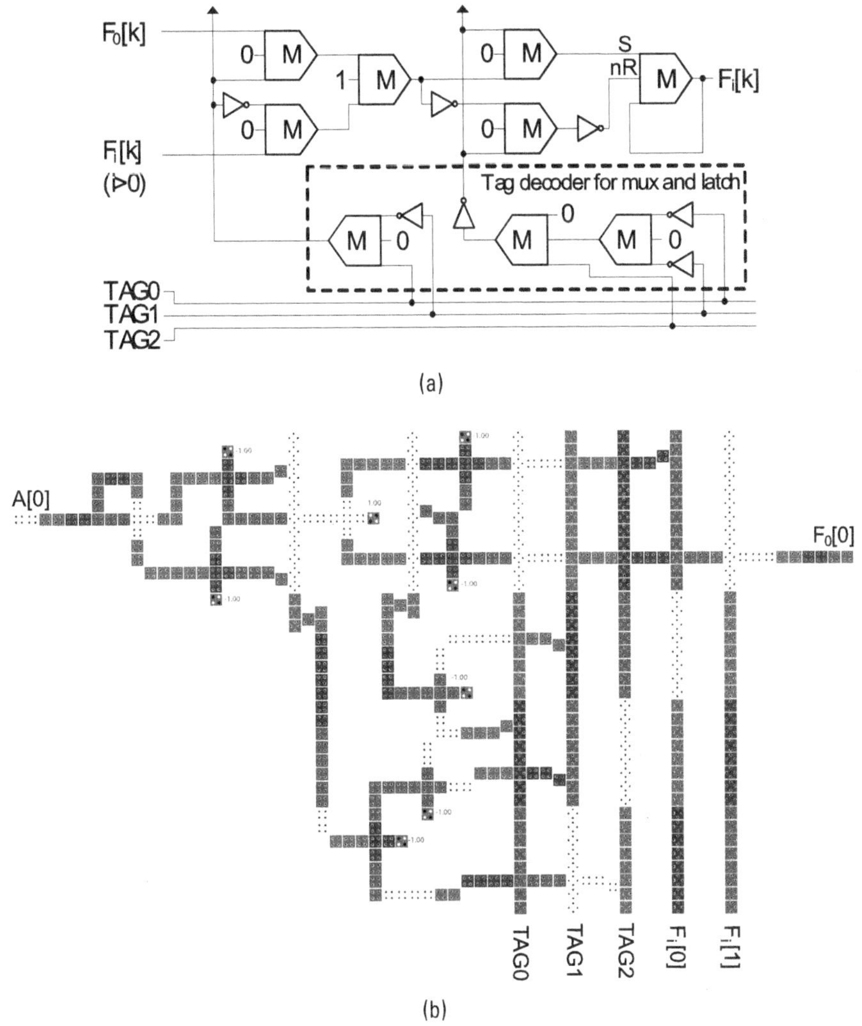

Figure 5.18 Tag decoder for F_i for the 24-bit Goldschmidt divider: (a) schematic, (b) layout for $k=0$. (© 2009 IEEE. *After* [9].)

5.3.3.6 Dividers of Other Sizes

To realize the 24-bit divider, much of the design remains the same as for the 12-bit divider. The multiplier size is increased to 24-bit by 24-bit (which increases its latency to 94 clock cycles). If the ROM for F_0 is kept as eight words of 3 bits, four iterations are needed to achieve an accuracy of 1 LSB. If the ROM size is increased to 32 words of 5 bits, only three iterations are needed. The larger ROM may have slightly larger latency, but since one iteration (that includes a pair of high latency multiplications) is saved, the total divider latency is reduced. Other elements of the 24-bit divider (for example, the latches used to

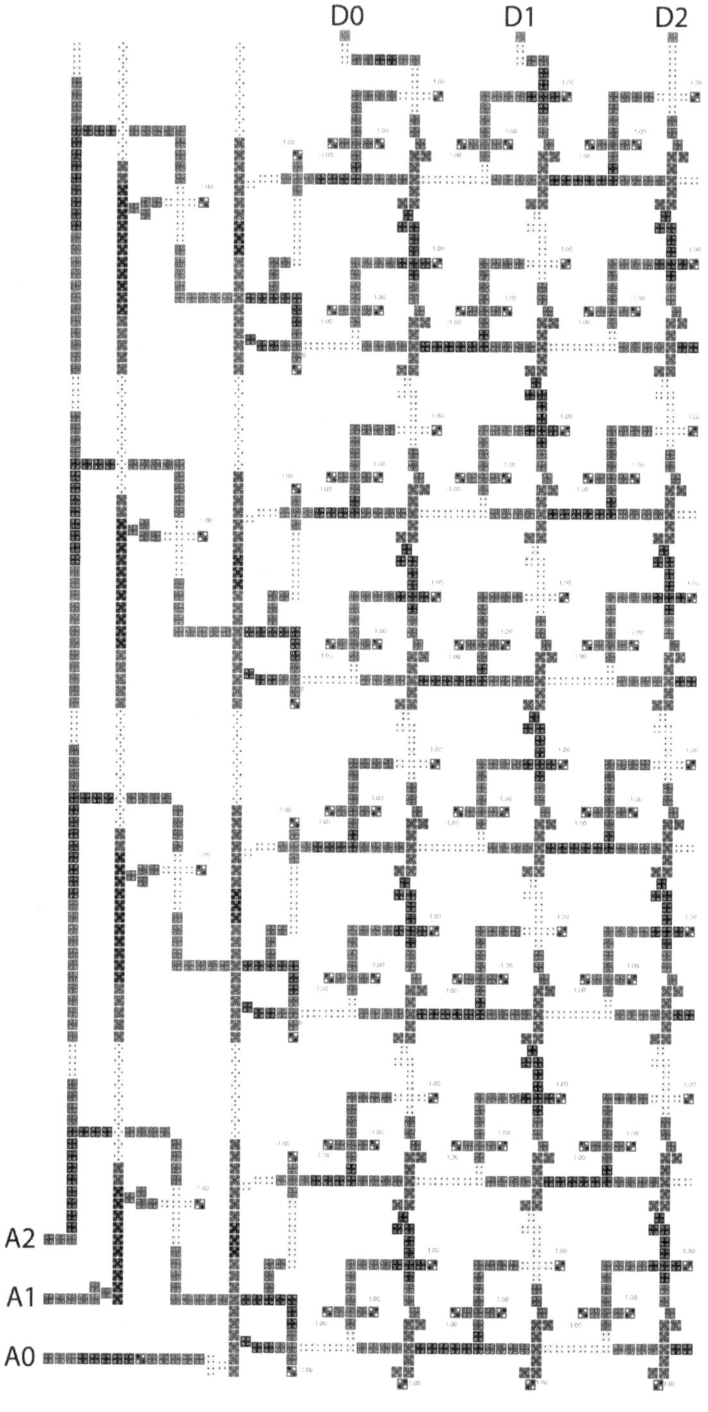

Figure 5.19 Layout of the eight-word by 3-bit reciprocal ROM. (© 2009 IEEE. *After* [9].)

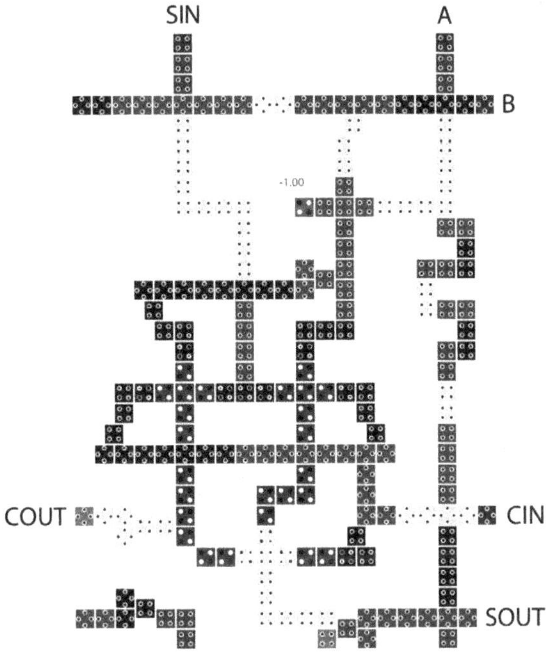

Figure 5.20 Layout of a cell for the array multiplier. (© 2009 IEEE. *From* [9].)

hold the data, and the word-wide inverter used to form the one's complement) will increase in size, but should not significantly impact the latency. Also the tag generator and decoders will increase in size as shown in Figures 5.14–5.15 and 5.17–5.18, respectively.

5.3.4 Simulation Results

5.3.4.1 12-bit Goldschmidt Divider

The layout of the 12-bit Goldschmidt divider is shown on Figure 5.22. The design has been implemented and simulated using QCADesigner [8]. In the simulations, most default parameters for a bistable approximation as listed in Section 2.4.1.3 are used. The number of samples is determined to be 226,000.

The area for the 12-bit Goldschmidt divider is 89.6 μm (8,818 nm × 10,158 nm), and the total number of QCA cells is 55,562. The delays of the functional units are shown in Table 5.2. Since a division requires three iterations, the total latency for a single isolated division is 219 clock cycles. Although this latency (in terms of the number of clock cycles) seems quite high, the clock rate for semiconductor QCA is on the order of one terahertz, so the time per isolated division can be less than 1/4 ns. Given the two clock cycles (one for *D* and the second for *N*) with a latency of 73 cycles per iteration, as

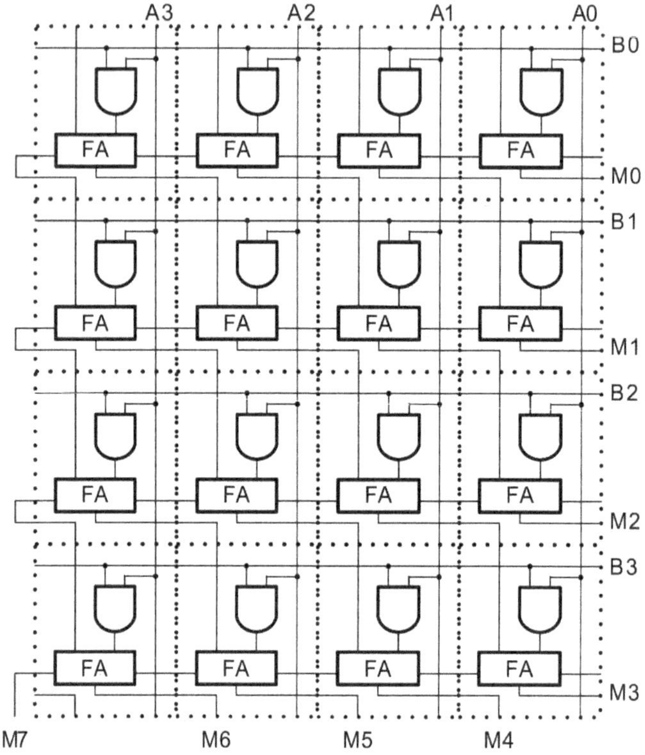

Figure 5.21 Schematic of a 4 × 4-bit multiplier using the multiplier cell.

many as 35 divisions can be started while the first one is progressing. Then successive quotients are available on every other clock cycle.

The 12-bit Goldschmidt divider was tested using bottom-up verification since a full simulation for a case takes about seven hours. Each unit block is verified exhaustively, and then the full integration is tested. A simulation of four consecutive divisions is shown in Figure 5.23. The first division computes 0.7080/0.7915. The inputs $D = 0.65516$ and $N = 0.5aa_{16}$ are shown at the left side of the second row of Figure 5.23(a). The results for this division ($Q = 0.894510$) are shown starting at clock 218 (sequence da_{16}) on the second row of Figure 5.23(b), D $= 0.7ff_{16}$ and $Q = N = 0.728_{16}$. Three additional divisions are performed immediately after the first division to show that pipelining achieves a peak division throughput of one division for every two clock cycles.

Table 5.3 shows the results for the four example divisions. The first two columns give the numerator and denominator in hexadecimal and decimal, respectively. Column 3 gives the exact quotient. Columns 4 and 5 give the result computed by the Goldschmidt divider in hexadecimal and decimal. Finally the last column gives the difference between the exact and the computed (QCA)

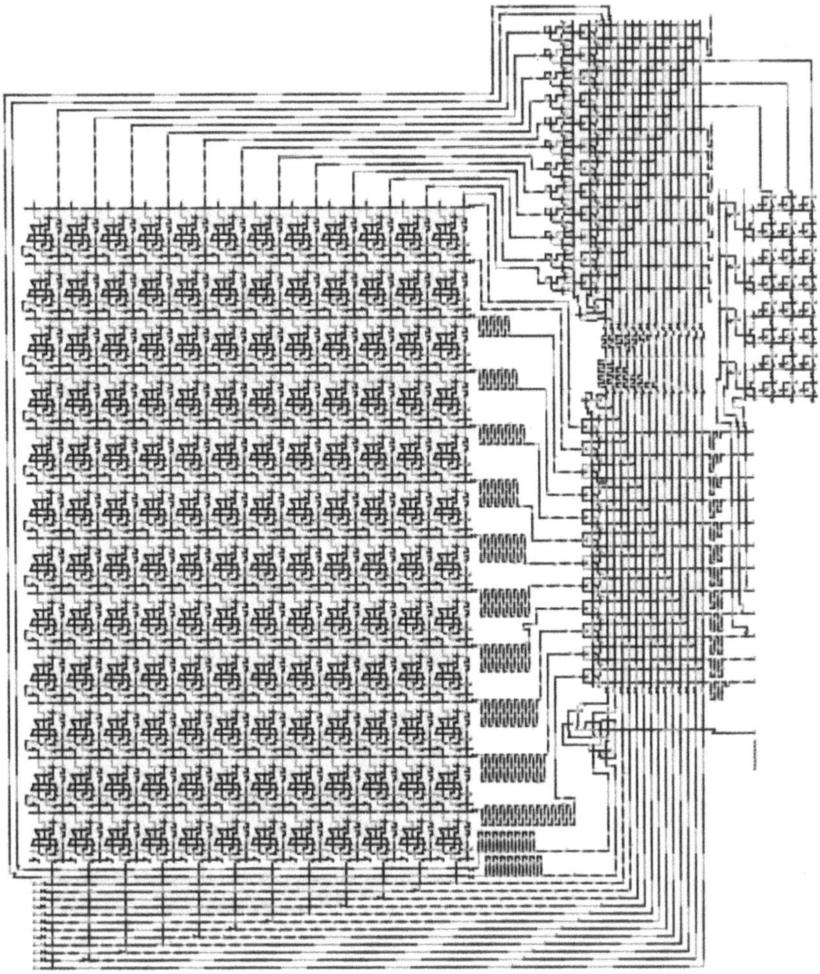

Figure 5.22 Layout of the 12-bit Goldschmidt divider. (© 2009 IEEE. *From* [9].)

Table 5.2
Delays of the Functional Units

Functional Unit	Delay (Clocks)
Tag generator	3
MUX and tag decoder	19
8 by 3 ROM	7
12-bit array multiplier	46
Data bus	5

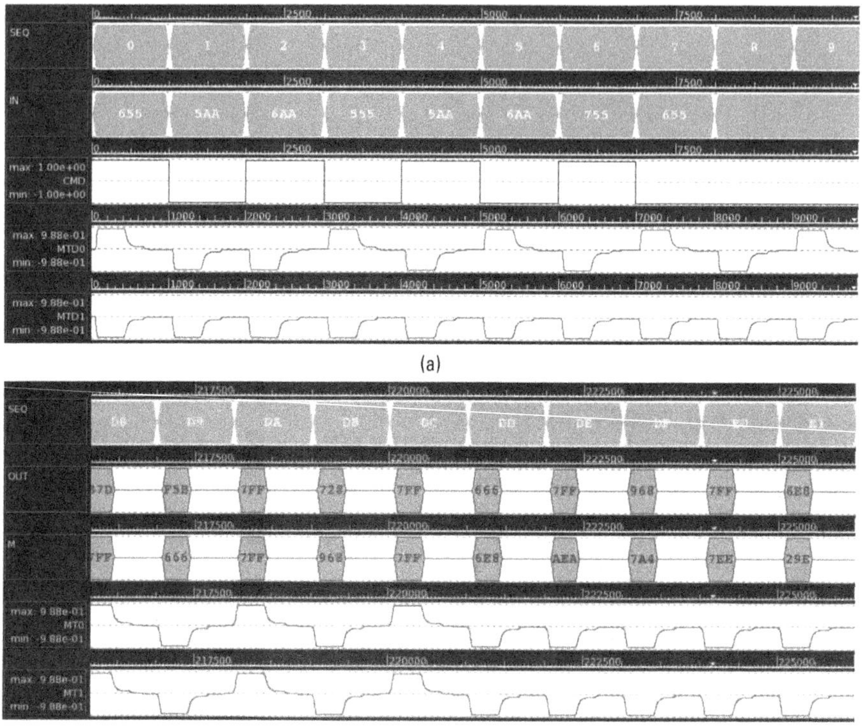

(a)

(b)

Figure 5.23 Simulation results: (a) input vectors for four consecutive divisions, (b) output waveforms for the four quotients. (© 2009 IEEE. *From* [9].)

quotients. In all four cases the computed quotients are accurate to within about 1 LSB.

5.3.4.2 24-bit Goldschmidt Divider

The layout of the 24-bit Goldschmidt divider is quite similar to the 12-bit divider, with the exception of the width of the data paths and the size of the array multiplier. The latency of all the elements except the multiplier (94 clock cycles) are as given in Table 5.2. Since four iterations are required, the latency for a single isolated division is expected to be about 484 clock cycles.

5.3.4.3 Data Tags for CMOS

Applying data tags to make a CMOS implementation of the Goldschmidt divider differs from the QCA version in that the computational circuits will generally have many fewer pipeline stages. For example, a 12-bit by 12-bit CMOS multiplier might be implemented with a Wallace or Dadda multiplier [21, 22] with only a few pipeline stages. In CMOS the lines that convey the data tags

Table 5.3
Example Divisions

Input		Quotient			
Hex	**Decimal**	**Exact**	**QCA**	**Decimal**	**Delta**
5aa/655	0.708 0/0.791 5	0.894 5	728	0.894 5	−0.000 02
555/6aa	0.66 5/0.833 0	0.800 1	666	0.799 8	0.000 3
6aa/5aa	0.833 0/0.708 0	1.176 5	968	1.175 7	0.000 8
655/755	0.791 5/0.916 5	0.863 6	6e8	0.863 2	0.000 3

are not inherently clocked as they are with QCAs. This requires the addition of pipeline latches for the data tags so that they stay synchronized with the data.

5.4 Conclusion

A restoring binary divider has been designed for implementation with QCA technology. The RAD is implemented with CFS cell blocks. It can be easily enlarged without long data connections, but it is large and slow due to the restoring algorithm. However, by using a pipelined parallel structure it has a good throughput.

A Goldschmidt divider (an iterative computational circuit) for QCA is implemented efficiently in a new architecture using data tags. The proposed data tag method avoids the synchronization problems that arise with conventional state machines in QCA due to the long delays between the state machines and the units to be controlled. In the proposed architecture, it is possible to start a new division at any iteration stage of a previous issued operation. As a result, the throughput is significantly increased since multiple division computations can be performed in a time-skewed manner using one iterative divider. Using the data tag method, 12-bit and 24-bit fixed-point Goldschmidt dividers can be implemented without synchronization problems.

References

[1] Kim, S., and E. Swartzlander Jr., "Restoring Divider Design for Quantum-Dot Cellular Automata," in *Proceedings of the 11th IEEE Conference on Nanotechnology,* 2011, pp. 1295–1300.

[2] Gardiner, A. B., and J. Hont, "Comparison of Restoring and Nonrestoring Cellular-Array Dividers," *Electronics Letters,* Vol. 7, No. 8, 1971, pp. 172–173.

[3] Parhami, B., *Computer Arithmetic Algorithms and Hardware Designs,* New York: Oxford University Press, Inc., 2000.

[4] Ercegovac, M. D., and T. Lang, *Division and Square Root: Digit-Recurrence Algorithms and Implementations,* Norwell, MA: Kluwer Academic Publishers, 1994.

[5] Kim, K., K. Wu, and R. Karri, "Towards Designing Robust QCA Architectures in the Presence of Sneak Noise Paths," in *Proceedings of the Conference on Design, Automation and Test in Europe–Volume 2,* 2005, pp. 1214–1219.

[6] Kim, K., K. Wu, and R. Karri, "The Robust QCA Adder Designs Using Composable QCA Building Blocks," *IEEE Transactions on Computer-Aided Design of Integrated Circuits and Systems,* Vol. 26, 2007, pp. 176–183.

[7] Liu, W., et al., "Design Rules for Quantum-Dot Cellular Automata," in *Proceedings of the IEEE International Symposium on Circuits and Systems,* 2011, pp. 2361–2364.

[8] Walus, K., et al., "QCADesigner: A Rapid Design and Simulation Tool for Quantum-dot Cellular Automata," *IEEE Transactions on Nanotechnology,* Vol. 3, 2004, pp. 26–31.

[9] Kong, I., E. Swartzlander, Jr., and S. Kim, "Design of A Goldschmidt Iterative Divider for Quantum-Dot Cellular Automata," in *Proceedings of the IEEE/ACM International Symposium on Nanoscale Architectures,* 2009, pp. 47–50.

[10] Walus, K., et al., "Simple 4-bit Processor Based on Quantum-dot Cellular Automata (QCA)," in *Proceedings of the 16th IEEE International Conference on Application-Specific Systems, Architecture Processors,* 2005, pp. 288–293.

[11] Goldschmidt, R. E., "Applications of Division by Convergence," Master's thesis, Massachusetts Institute of Technology, 1964.

[12] Obermann, S. F., and M. J. Flynn, "Division Algorithms and Implementations," *IEEE Transactions on Computers,* Vol. 46, 1997, pp. 833–854.

[13] Richards, R., *Arithmetic Operations in Digital Computers,* New York: D. Van Nostrand Co., 1955.

[14] Anderson, S., et al., "The IBM System/360 Model 91: Floating-Point Execution Unit," *IBM Journal of Research and Development,* Vol. 11, No. 1, 1967, pp. 34–53.

[15] Zhang, R., et al., "A Method of Majority Logic Reduction for Quantum Cellular Automata," *IEEE Transactions on Nanotechnology,* Vol. 3, 2004, pp. 443–450.

[16] Walus, K., et al., "Circuit Design Based on Majority Gates for Applications with Quantum-dot Cellular Automata," in *Conference Record of the 38th Asilomar Conference on Signals, Systems and Computers,* Vol. 2, 2004, pp. 1354–1357.

[17] Huang, J., M. Momenzadeh, and F. Lombardi, "Design of Sequential Circuits by Quantum-dot Cellular Automata," *Microelectronics Journal,* Vol. 38, No. 4-5, 2007, pp. 525–537.

[18] Kim, S., and E. Swartzlander, Jr., "Parallel Multipliers for Quantum-dot Cellular Automata," in *Proceedings of the IEEE Nanotechnology Materials and Devices Conference,* 2009, pp. 68–72.

[19] Hänninen, I., and J. Takala, "Pipelined Array Multiplier Based on Quantum-dot Cellular Automata," in *Proceedings of the 18th European Conference on Circuit Theory and Design,* 2007, pp. 938–941.

[20] Cho, H., and E. Swartzlander, Jr., "Adder and Multiplier Design in Quantum-dot Cellular Automata," *IEEE Transactions on Computers,* Vol. 58, 2009, pp. 721–727.

[21] Wallace, C. S., "A Suggestion for a Fast Multiplier," *IEEE Transactions on Electronic Computers,* Vol. EC-13, 1964, pp. 14–17.

[22] Dadda, L., "Some Schems for Parallel Multipliers," *Alta Frequenza,* Vol. 34, 1965, pp. 349–356.

Part III
QCA Design Methodologies

6

Design of QCA Circuits Using Cut-Set Retiming[1]

Weiqiang Liu, Liang Lu, Máire O'Neill, Earl E. Swartzlander, Jr., and Roger Woods

6.1 Introduction

To design QCA circuits, an obvious approach is to map existing CMOS designs into their majority logic counterparts. However, the design of a QCA circuit is radically different from a conventional digital design due to timing considerations and requires different design optimization techniques. Characteristics that are specific to QCA need to be taken into account including its four-phase clocking, which enables deep pipelines, and the unique "layout=timing" problem [3]. Although QCA logic components can be designed with majority gates, extra delays inserted into QCA wires can lead to incorrect timing relationships producing incorrect results. Therefore, assigning a correct and efficient clocking scheme becomes very important. Furthermore, as there are timing constraints on wires in QCA, the implementation of feedback is also a critical problem.

To date, there has been little research on the timing issues of QCA. The "layout=timing" problem was recognized by Niemier and Kogge [3], who proposed a specific arrangement of clocking—namely, trapezoidal clocking—to reduce the design area. This was also considered as a possible method to implement feedback paths. Two-dimension clocking was proposed by Vankamamidi et al. [4] to reduce the longest line length in each clocking zone. However, none of this previous work provides any practical design of feedback. Based on conventional RS-type and D-type flip-flops, a stretching algorithm for assigning clocking zones to QCA sequential circuits by matching delays was proposed

1. Chapter 6 is based on [1, 2].

by Huang et al. [5]. Their proposed delay-matching design method ensures that all paths from the outputs to the inputs of flip-flops have the same delays. However, many unnecessary delays are introduced due to the strict matching strategy, resulting in an expansion of the overall number of cells and circuit size.

This chapter presents a cut-set retiming design procedure for QCA circuit design based on QCA timing constraints to solve the timing issues, which are further discussed in the Section 6.2. It utilizes features such as synchronization, deep pipelines, and local interconnection, which are common to both QCA and systolic architectures [6]. The proposed cut-set retiming design procedure described in this chapter can accommodate QCA characteristics and resolve the timing problems by performing time-scaling and delay-transfer to relocate the existing delays while keeping the timing relationship unchanged. The cut-set retiming can be performed to assign proper clocking zones and solve the feedback problem in QCA designs. Systolic and nonsystolic architectures are designed to illustrate the procedure. The design results show that complex circuits, such as a Montgomery modular multiplier can be designed in QCA, by using the proposed procedure.

The rest of the chapter is organized as follows. Section 6.2 discusses the QCA timing constraints and issues. Both CMOS and QCA data-flow graphs are introduced in addition to the retiming technique in Section 6.3. Section 6.4 proposes a cut-set retiming design procedure for QCA circuits. Section 6.5 presents and analyzes two case studies using the procedure, and Section 6.6 provides conclusions.

6.2 QCA Timing Constraints and Timing Issues

The new computation paradigm offered by QCA and its unique clocking scheme impose two basic timing constraints on QCA circuits. The first one is the minimum delay in QCA logic units, and the second one is the maximum number of cells in a single clocking zone. These two constraints must be met in QCA circuits to achieve correct functionality. Meanwhile, timing issues that arise due to the timing constraints, such as heavy interconnection latency and the feedback problem, make the timing design very difficult in QCA circuit design.

6.2.1 Timing Constraint I

In QCA circuits, even if the logic is equivalent to that of a CMOS circuit, synchronization may be difficult. As discussed in Section 2.6, the timing constraint on a QCA majority gate is that all three inputs must reach the device cell (central cell) at the same time in order to have correct operation. If all three input wires are equally long, the device cell can be within the same clocking zone as the inputs. However, in practice, the lengths of the input wires are usually different. Therefore, these three inputs should be designed to be in the same clocking zone i, and the majority gate as well as its output should be in the subsequent clocking zone $[(i+1) \bmod 4]$ to achieve fair voting. As a result, at least one clocking zone delay (denoted as D^{-1}) is required for a majority gate. The minimum delay for its derivatives, such as OR and AND gates, is also one clocking zone delay (D^{-1}). However, the QCA inverter has only one input that does not need to be synchronized; therefore one clocking zone is sufficient for QCA inverters. As most QCA logic units are constructed from majority gates, delays are inevitable. The minimum delay of a particular logic unit is determined by the number of cascaded majority gates. As mentioned in Section 2.6.2, although this timing rule comes from the simulations using the ICHA method, it should be considered to make a robust design in QCADesigner.

6.2.2 Timing Constraint II

Long QCA wires can result in an increased delay in signal propagation and switching, which can significantly reduce the overall operating speed. Therefore, the clock rate is improved if each clocking zone contains a small number of cells. The length of a QCA wire in a clocking zone is limited due to thermodynamic effects [7]. As a result, long QCA wires have to be divided into multiple clocking zones as shown in Figure 6.1 to ensure robust signal transmission. As a result, extra delays are incurred due to the added clocking zones.

Previous designs in QCADesigner [8, 9] show that a maximum length (denoted as L) of 25 cells is a reasonable assumption. However, the true maximum number of cells in QCA clocking zones depends on the fabrication technology.

6.2.3 Timing Issues in QCA

From the above discussion, it is clear that two basic timing constraints have to be met in QCA designs. These constraints present several timing issues, such

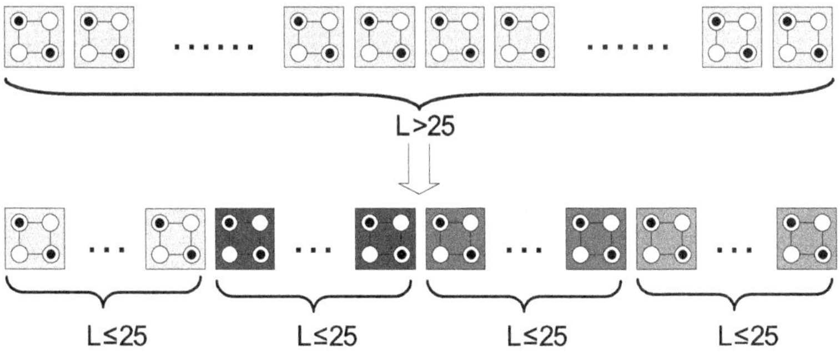

Figure 6.1 A long QCA wire partitioned into different clocking zones. (© 2011 IEEE. *From* [2].)

as proper clocking assignment, interconnection latency, and feedback difficulty. The success of designing efficient QCA circuits is heavily dependent on how these issues are dealt with.

6.2.3.1 Proper Clocking Assignment

In timing constraint I, all the inputs of a majority gate should arrive at the central cell at the same time, and the interconnection wires between two majority gates require proper clocking zone assignment to ensure synchronization. It is easy to assign clocking zones to a signal-forward architecture in QCA, as this only requires adding delays. However, for sequential circuits, especially those with feedback, the clocking zone assignment is not an easy task.

6.2.3.2 Interconnection Latency

As cells in a long QCA wire are partitioned into different clocking zones, and each clocking zone can be considered as a latch with one clocking zone delay (D^{-1}), significant interconnection latency is introduced by timing constraint II. Meanwhile, QCA logic units composed of majority gates contribute extra latency to the circuit. As a result, the latency is determined by the largest number of clocking zones between the inputs and the outputs. The latency that exists in QCA wires and majority gates should be carefully considered when mapping a CMOS architecture into its QCA counterpart. Minimizing the latency (similar to reducing the critical path length in CMOS) is a major factor in achieving an efficient QCA circuit design.

6.2.3.3 Feedback Problem

The ability to handle feedback paths is another issue arising from the two timing constraints. Feedback in CMOS can be easily implemented by physical wires connected from outputs to inputs. However, in QCA there is no physical wire. All the QCA wires are composed of coupled cells within particular clocking zones. Therefore, a long feedback path will traverse a number of clocking zones. This may lead to asynchrony problems with the next input signal. For example, as shown in Figure 6.2 (U_i denotes the ith logic unit and D^{-di} denotes its delays), if U_0 produces a signal that should be connected to $U1$, U_2, ..., U_k, a straightforward approach is to connect them using QCA wires within

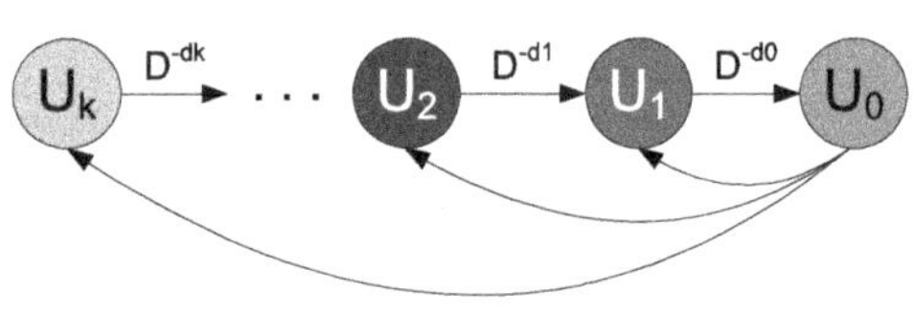

Figure 6.2 An example of the feedback problem in QCA. (© 2011 IEEE. *From* [2].)

one clocking zone to make sure the feedback arrives at $U_i,(i = 1, 2, \ldots, k)$at the same time. However, the feedback from U_0 to U_k (assuming that k is large) could require a long QCA wire with a large number of QCA cells. According to the second timing constraint, this is not permitted. Long QCA wires should be divided into multiple clocking zones, which introduces undesired delays along the wire. The timing relationship is changed due to the extra delays. Designing this kind of feedback in QCA is a significant issue.

6.3 Data Flow Graph and Retiming Technique

6.3.1 Data Flow Graph

The data flow graph (DFG) is a graphic representation approach for algorithms. In DFG representations, the nodes represent computations and the directed edges represent data paths (communications between nodes) and each edge has a nonnegative number of delays associated with it [10]. Figure 6.3(b) is a DFG of the computation $y(n)= ay(n–1) + x(n)$, where node U represents addition and node V represents multiplication, the edge from U to V contains one delay and the edge from V to U contains no delay. Associated with each node is its execution time in terms of normalized time units (*u.t.*). For example, the execution time of node U is 2 *u.t.* and the execution time of node V is 4 *u.t.* as shown in Figure 6.3(b).

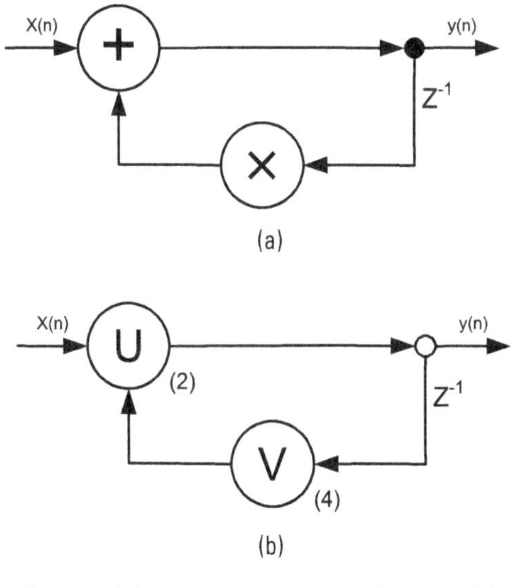

Figure 6.3 (a) Block diagram of the computation $y(n)= ay(n-1) + x(n)$ and (b) its DFG representation. (© 2011 IEEE. *From* [2].)

6.3.2 Mapping CMOS DFG to QCA DFG

As discussed above, in the conventional DFG representation, the nodes represent computations (or functions) with their execution time in terms of normalized time units, and the directed edges represent data paths with a nonnegative number of delays associated with them [10]. When a DFG is applied to a QCA circuit, as the execution time is measured in terms of clocking zone delays, the computation times of the nodes are put onto the outbound edges as delays.

For example, the DFG of $y(n) = ay(n-1) + x(n)$ is shown again in a simplified way in Figure 6.4(a). Assuming the execution of computations U and V require two clocking zone delays (D^{-2}) and four clocking zone delays (D^{-4}), respectively, the CMOS DFG can be mapped into Figure 6.4(b). In the QCA DFG, the execution delays are put into the outbound edges, as shown in Figure 6.4(c). This QCA DFG representation is used in the rest of this chapter.

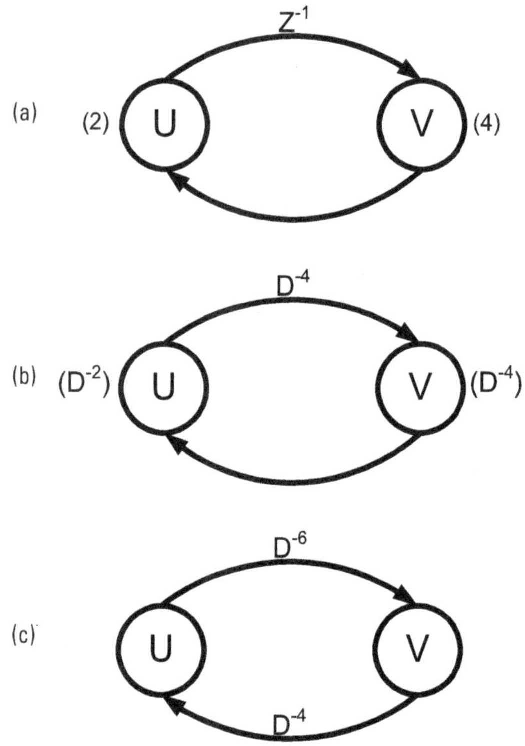

Figure 6.4 Computation of $y(n) = ay(n-1) + x(n)$: (a) conventional CMOS DFG representation, (b) converted representation using $Z^{-1} = D^{-4}$ in QCA, (c) QCA DFG representation.

6.3.3 Retiming Technique

Retiming is a transformation technique used to change the locations of delay elements such as registers in circuits to improve the performance, area, or power characteristics without affecting the input and output relationship of the circuit [11]. The central objective of retiming in CMOS technology is to design a circuit with the minimum number of registers. In this chapter, retiming techniques are applied to QCA circuits under the two QCA timing constraints. This technique resolves QCA timing issues and achieves efficient QCA designs in terms of latency and area.

The retiming technique uses a DFG with the vertices representing asynchronous combinational blocks and the directed edges representing interconnection with registers. Each vertex has a value that indicates the delay through the combinational circuit it represents. Two basic transformations are used to achieve retiming. One is deleting a register from each input of a vertex while adding a register to all outputs. The other is the converse, adding a register to each input of a vertex and deleting a register from all outputs.

Given a DFG, $G := (V, e)$, where V indicates vertices (logic units) and e represents edges, a retiming solution is characterized by a value $r(V)$ for each node V in the graph. Let $w(e)$ denote the weight of edge e in the original graph and let $w_r(e)$ denote the weight of edge e in the retimed graph. As shown in Figure 6.5, the objective of retiming is to compute a nonnegative integer lag value $r(V)$ from each vertex U to every other vertex V using:

$$w_r(e) = w(e) + r(V) - r(U) \tag{6.1}$$

Automatic algorithms are available to find the retiming values [12]. The retiming only changes the weight of the edges (delay elements) without affecting the functionality of the circuit.

6.4 A Cut-Set Retiming Design Procedure

6.4.1 Cut-Set Retiming and Its Rules

Cut-set retiming, a special case of retiming, is a graphical technique with the advantage of solving complex retiming problems in a simple way. Cut-set reti-

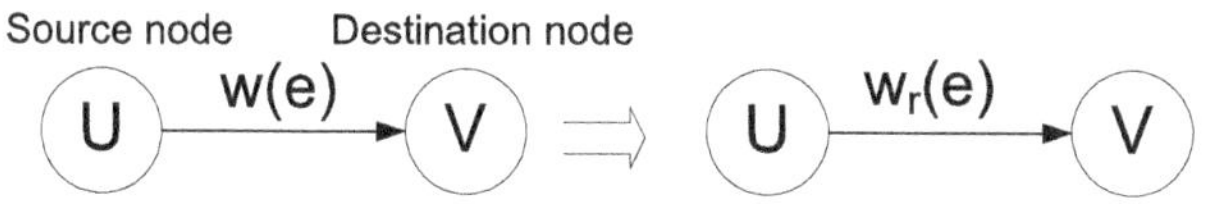

Figure 6.5 Retiming formulation. (© 2011 IEEE. *After* [2].)

ming was originally proposed as a general systematic design method to generate systolic arrays [6]. Since QCA and systolic architectures are similar in terms of synchrony, deep pipelines, and local interconnection, the success of cut-set retiming with systolic arrays implies the feasibility to resolve the timing issues in QCA using this technique. A cut-set is defined as a minimal set of edges that can be removed from the graph to create two disconnected subgraphs. Cut-set retiming only affects the weights of the edges in the cut-set.

There are two basic rules for cut-set retiming: time-scaling and delay-transfer.

6.4.1.1 Time-Scaling

To accommodate combinational logic delays and the wire length limitation, more delays may be needed than those that already exist. The number of delays present on the edges of a graph may be scaled by a single positive integer without altering the overall timing. All the delays may be scaled up by a scaling factor (denoted as α) due to the two basic QCA timing constraints discussed above. As illustrated in Figure 6.6, delays can be scaled up by a scaling factor of α, which implies that the operation can be performed once every α clock cycles. The scaling factor is determined by the worst-case loop bound, as defined by (6.2):

$$\alpha = \frac{R_D}{E_D} = \frac{R_{D1} + R_{D2}}{E_D} \tag{6.2}$$

where, E_D refers to the delays that already exist, and R_D refers to the delays required in the worst case loop by the two timing constraints, which can be divided into two parts: minimum delays required by logic units (R_{D1}) and delays introduced in the QCA wires (R_{D2}). However, there is no simple way to estimate the length of wires from the DFG. After time-scaling [6], there should be sufficient delays for transferral between edges. Note that time-scaling is also

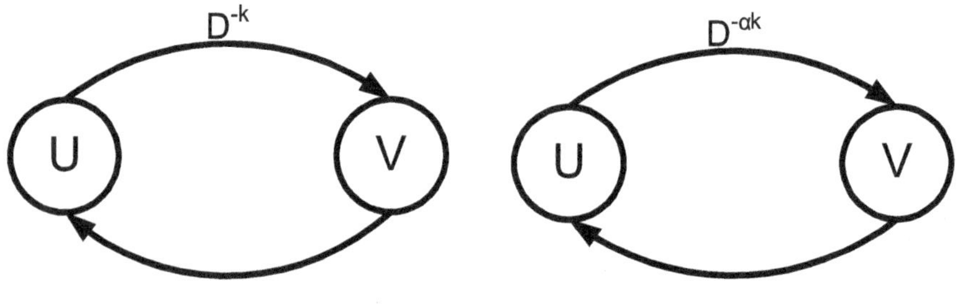

Figure 6.6 Time-scaling demonstration: all existing delays scaled up by α. (© 2011 IEEE. *From* [2].)

known as slow down [10]. In an α-slow system, $\alpha - 1$ null operations (or 0 samples) must be interleaved after each useful signal sample to preserve the functionality of the algorithm.

6.4.1.2 Delay-Transfer

Cut-set retiming is achieved by adding *K* delays to inbound edges and removing *K* delays from the outbound edges, and vice versa. As shown in Figure 6.7(a), cut-set retiming is performed by adding *K* delays to each edge from *V* to *U* and removing *K* delays from each edge from *U* to *V*. This is referred to as delay-transfer [6]. In a special case as shown in Figure 6.7(b), delays can be directly inserted into forward edges. Cut-set retiming only affects the edges in the cut-set. As a result, the input-input and input-output timing relationships remain the same.

6.4.2 Proposed Cut-Set Retiming Design Procedure

To resolve the timing issues and achieve efficient QCA circuit design, cut-set retiming is proposed subject to the two QCA timing constraints. The design procedure based on the time-scaling and delay-transfer rules is outlined as follows:

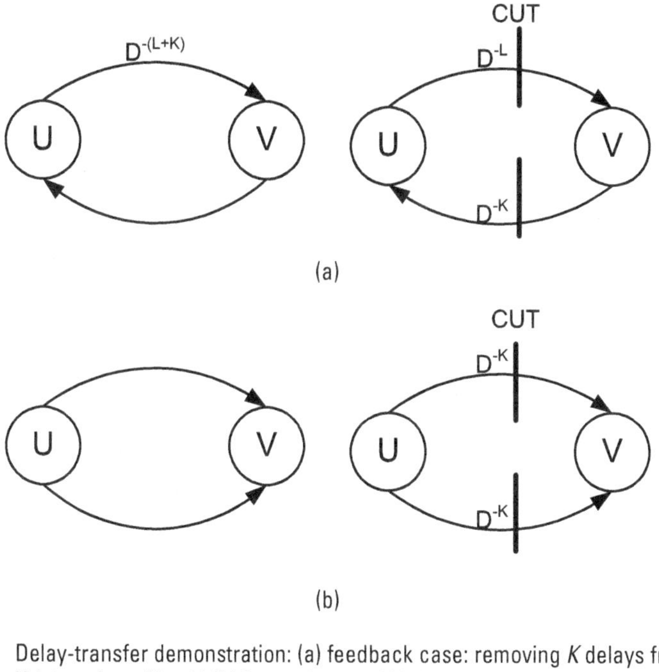

Figure 6.7 Delay-transfer demonstration: (a) feedback case: removing *K* delays from *U* to *V* and adding *K* delays from *V* to *U*; (b) forward case: *K* delays can be added to edges directly. (© 2011 IEEE. *After* [2].)

- Step 1: Map CMOS architecture to QCA architecture using the QCA DFG.
- Step 2: Majority gate based logic optimization: All the logic units should be optimized by using the minimum number of majority gates [13].
- Step 3: Determine the delays of the logic units: The selection of logic units may not be unique. Different choices may lead to different delays. Once the basic units are selected, their minimum delays can be determined by timing constraint I (i.e., the number of cascaded majority gates).
- Step 4: Timing analysis: The existing delays in the original architecture may be insufficient for transferral. Therefore, timing analysis is necessary. The scaling factor (α) relies on the worst case loop and can be calculated by (6.2). If $\alpha \leq 1$, then there is no requirement for time scaling. If $\alpha > 1$ then all existing delays should be scaled up by $\lceil \alpha \rceil$. The delays introduced by long wires cannot be determined immediately until the layout is designed. Therefore, estimated delays are used before the actual layout is completed and a further evaluation is required post-layout. As a result, the timing analysis may need to be performed more than one time.
- Step 5: Insert and transfer delays: Delays can be inserted to signal-forward edges that are not in a loop with feedback. Existing delays are transferred by the delay-transfer rule according to minimum delays of the logic units as well as delays introduced in long wires.
- Step 6: Layout and evaluation of long wires: Layouts can be mapped from the retimed architectures. Wire delays post-layout are evaluated as the layout may produce some new long wires that cannot be determined at the architecture design stage. If there are any wires exceeding the length limitation, the timing should be reanalyzed and steps 3 and 4 repeated.
- Step 7: Verification: When all the timing constraints are satisfied, the functionality of the designed layout can be verified by using QCADesigner [14].

A flow chart of the cut-set retiming design procedure is shown in Figure 6.8.

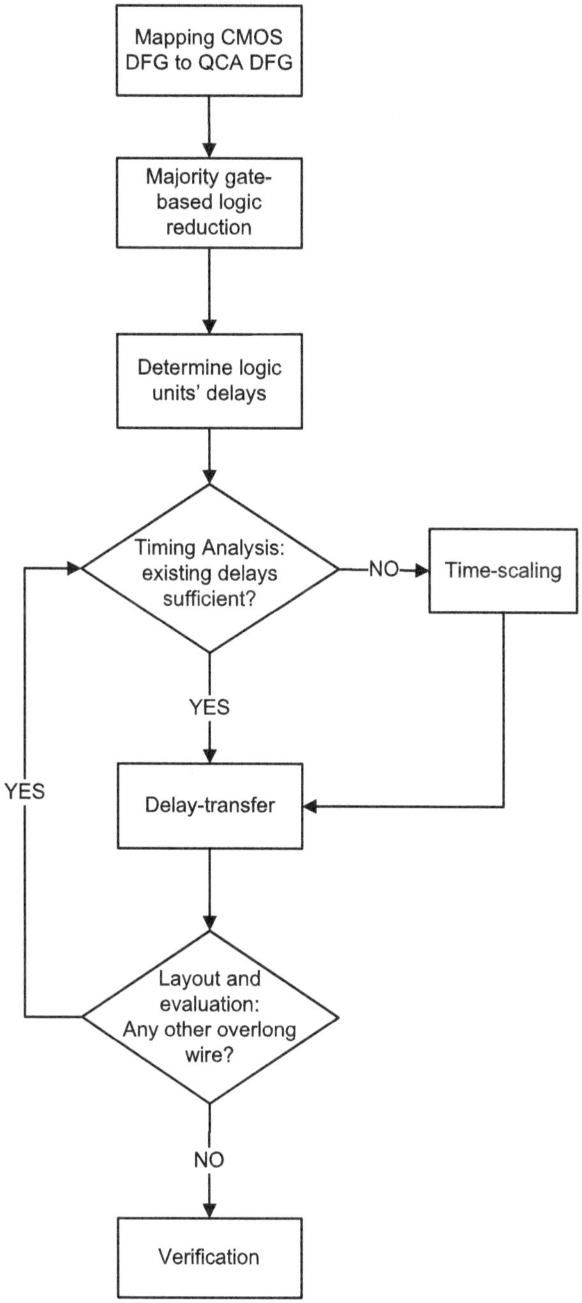

Figure 6.8 Flow chart of the cut-set retiming design procedure. (© 2011 IEEE. *After* [2].)

6.5 Case Studies

To illustrate the proposed design procedure, systolic array architectures are considered since the cut-set retiming technique was originally employed in systolic design. In this section, both a systolic architecture and a nonsystolic architecture are designed. A systolic Montgomery modular multiplier (MMM) architecture is studied first. Then a nonsystolic architecture, which is a benchmark circuit (ISCAS'89 S27 [15]), is designed and compared with an equivalent architecture by Huang et al. [5], which was designed with a delay-matching design method. These examples show that the proposed cut-set retiming design procedure is easily applied to both systolic and nonsystolic architectures to achieve more efficient designs.

6.5.1 MMM Design

Modular multiplication is a core operation in most public-key encryption schemes such as RSA [16] and elliptic curve cryptography (ECC) [17]. MMM is the most useful and efficient method for performing a fast modular multiplication in hardware [18]. In each iteration, the Montgomery algorithm performs an addition operation and a shift operation with the least significant bits (LSB) of the intermediate results. The Montgomery algorithm computes $MonPro(A, B, M) = A \times B \times R^{-1}$ modM, where A and B are two positive integers, (i.e., $0 < A, B < M$); R is usually taken to be a power of 2 in order to perform fast division by right shift operations; and M is the modulus. The algorithm requires that $gcd(M, R) = 1$.

The Montgomery algorithm has been modified for systolic implementation [19]. A 4-bit systolic MMM architecture is represented by the CMOS DFG shown in Figure 6.9, where, "+" indicates a FA and the circles with A_i and $M_i'(M' = \frac{M+1}{2})$ indicate AND operations. The circles with B_i and S_i indicate the serial input and intermediate result, respectively. The MMM architecture is difficult to design in QCA due to the long wires associated with the inputs and the feedback.

Furthermore, there is a serial adder chain, which makes the timing requirement difficult to achieve.

An efficient QCA MMM architecture can be achieved using the cut-set retiming procedure as follows:

- Step 1: The CMOS DFG is mapped into the QCA DFG using $Z^{-1} = D^{-4}$, as shown in Figure 6.10.
- Step 2: The logic units in this architecture are logic ANDs and serial adders. The AND gate is implemented in QCA by a majority gate with

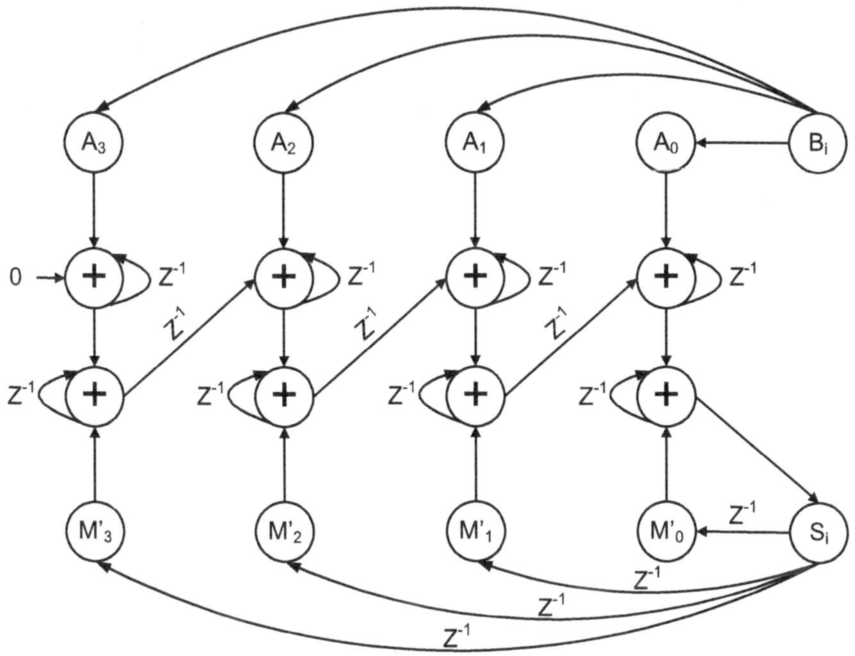

Figure 6.9 CMOS DFG of a four-bit serial systolic MMM architecture.

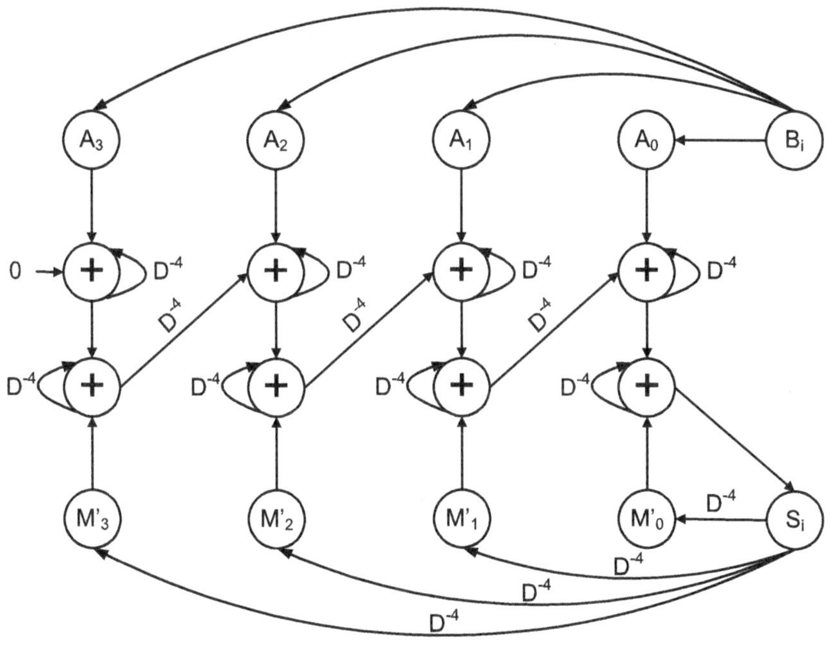

Figure 6.10 QCA DFG of a four-bit serial systolic MMM architecture. (© 2011 IEEE. *From* [2].)

one input fixed to "0." The adder design can be optimized by using just three majority gates [13].

- Step 3: The AND gates require one clocking zone delay (D^{-1}) as discussed in Section 6.2. The serial adder used in this design was proposed by Cho and Swartzlander [20], and there are two cascaded majority gates from the inputs to the outputs. Therefore, the bit-serial adder requires two clocking zone delays (D^{-2}).
- Step 4: Timing analysis is required to make sure there are sufficient delays. The worst case loop in this architecture is shown in Figure 6.11. In this loop, the number of existing delays (E_D) is 16 D^{-1}. The required delays (R_D) depend on the minimum delays of the logic units and the delays in the long wires. Considering R_{D1}, the minimum delay required by the logic units, there are seven serial adders and one AND gate in the loop. According to step 2, these units require $7 \times 2 + 1 = 15$ clocking zone delays, (i.e., $R_{D1} = 15D^{-1}$). Next, considering an estimate for R_{D2}, the QCA wire delay, feedback wires from S_i to M'_3 travel over three serial adders (with 0 a similar situation for the input wire from B_i to A_3), and each serial adder has a length of 18 cells from the inputs to the outputs [20]. Therefore, the long feedback wire will travel over more than

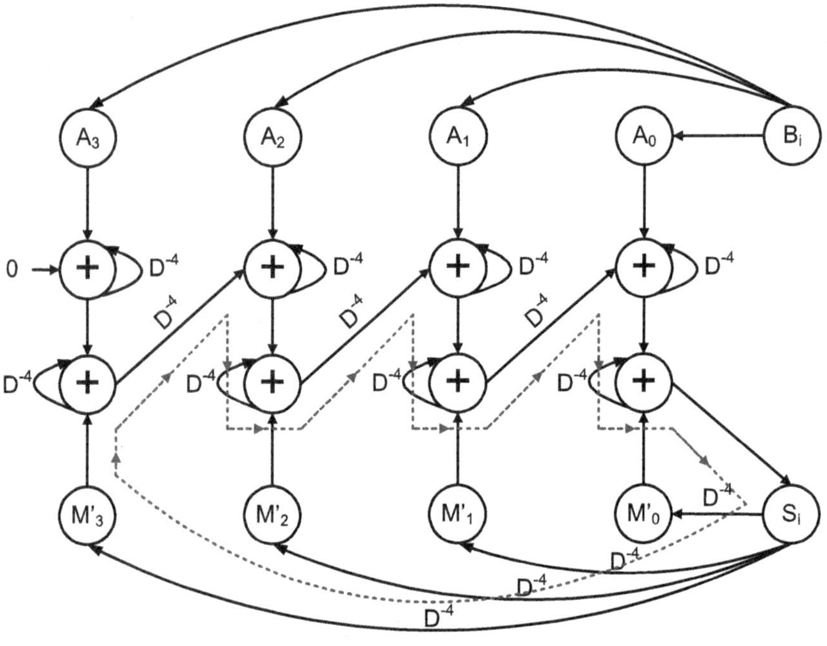

Figure 6.11 Timing analysis: the worst-case loop in the MMM architecture. (© 2011 IEEE. *From* [2].)

$3 \times 18 = 54$ cells, which exceeds the predefined length limitation of 25. The feedback wire should be evenly divided into different clocking zones. Therefore, the R_{D2} delay needs to be more than $1D^{-1}$ but less than $16\ D^{-1}$. As a result, the scaling factor is calculated according to Equation (6.2) as follows:

$$\alpha = \frac{R_D}{E_D} = \frac{R_{D1} + R_{D2}}{E_D} = \frac{15 + R_{D2}}{16}$$
$$1 < R_{D2} < 16, \text{ therefore, } 1 < \alpha < 2$$

Thus, the existing delay is not sufficient, and time-scaling is required in this case. All the delays in Figure 6.11 should be scaled up by a scaling factor of $\lceil \alpha \rceil = 2$, which means two clock cycle delays ($Z^{-2} = D^{-8}$).

- Step 5: Delays can be inserted and transferred as illustrated in the 14 cuts shown from Figure 6.12–6.14. The fully retimed MMM architecture is shown in Figure 6.14.
- Step 6: The retimed architecture can now be easily mapped to a QCA MMM layout, as shown in Figure 6.15. In this layout, there is no wire

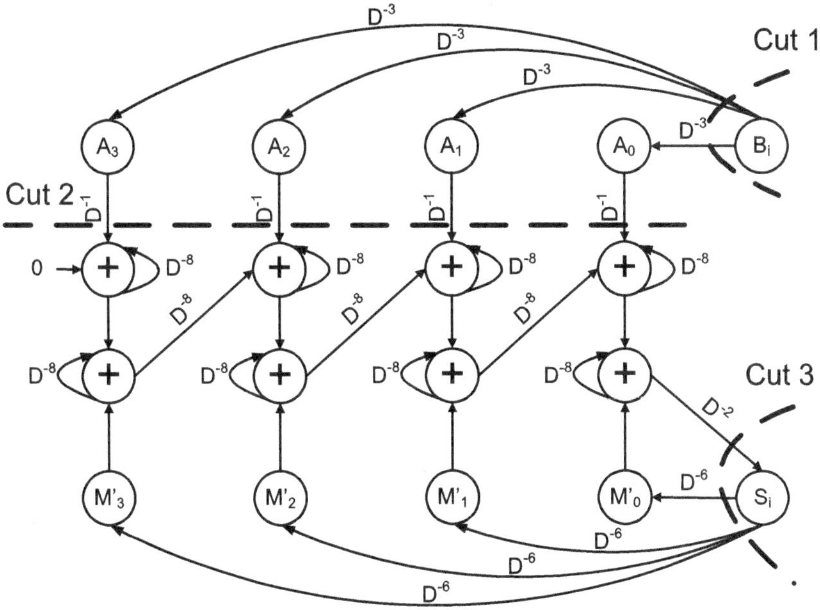

Figure 6.12 The cut-set retiming procedure for a 4-bit QCA MMM: cut 1 to cut 3. (© 2011 IEEE. *From* [2].)

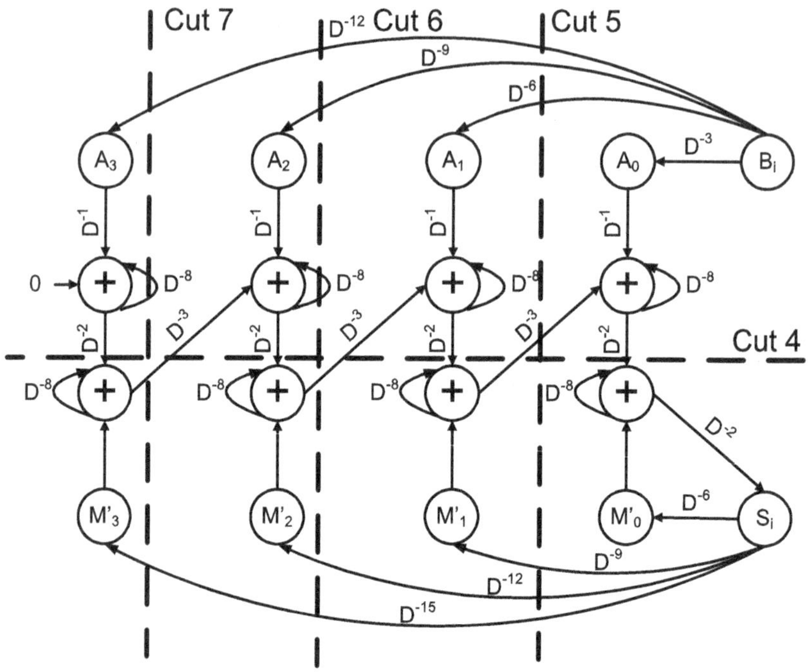

Figure 6.13 The cut-set retiming procedure for a 4-bit QCA MMM: cut 4 to cut 7. (© 2011 IEEE. *From* [2].)

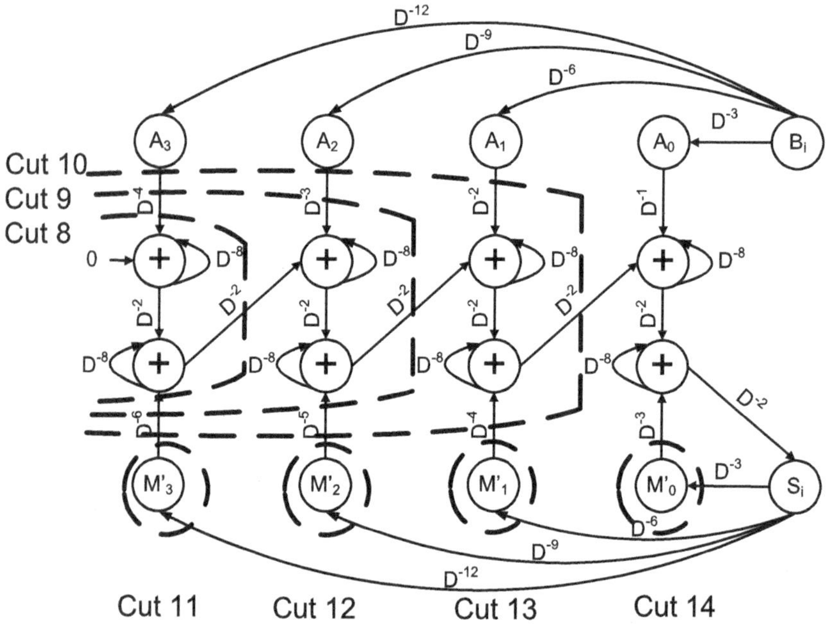

Figure 6.14 The cut-set retiming procedure for a 4-bit QCA MMM: cut 8 to cut 14. (© 2011 IEEE. *From* [2].)

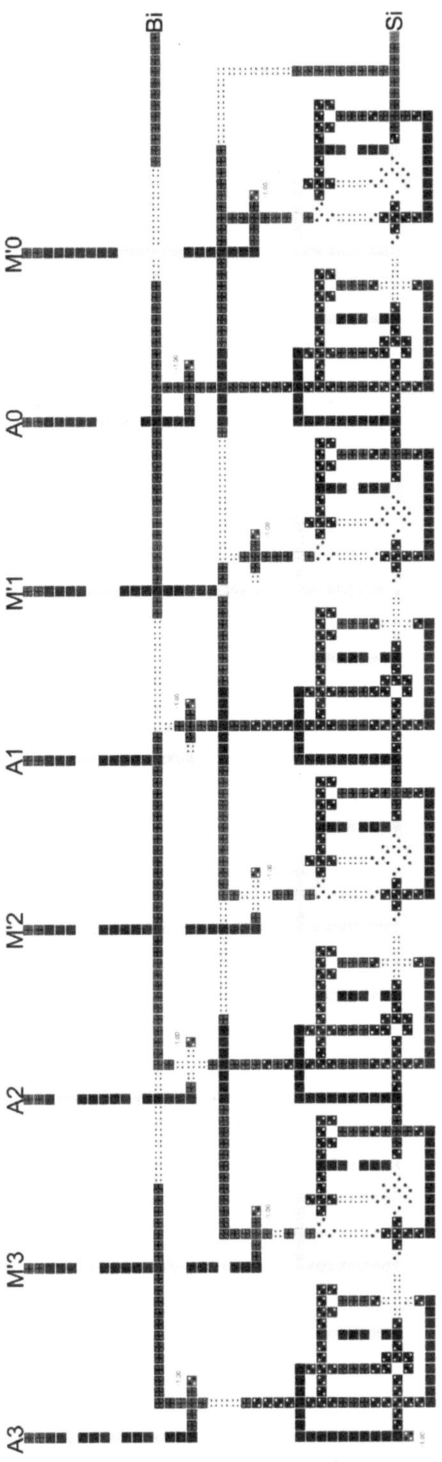

Figure 6.15 Layout of a 4-bit QCA Montgomery modular multiplier. (© 2011 IEEE. *From* [2].)

exceeding 25 cells in length. Thus, all timing constraints are met. However, if the future fabrication technology only permits fewer cells in a clocking zone, for instance 15 cells, then a further timing analysis is required for this QCA MMM architecture. From the layout, it can be seen that the cells in the wire from the AND gate associated with A_0 to the lower serial adder is more than 15 cells. In this case, one more clocking zone delay (D^{-1}) can be inserted in Cut 2 in Figure 6.12 to meet the new timing constraint. Then all subsequent steps are similar to that described above.

- Step 7: The 4-bit QCA MMM architecture was verified using QCADesigner [1]. The overall delay for a MMM with n-bit inputs is $2n + 2$ clock cycles. It achieves a constant throughput (one bit per clock cycle) as long as the pipeline is kept full.

This is the first design of a MMM in QCA. It shows that circuit designs with complex timing requirements are possible.

6.5.2 S27 Benchmark Circuit Design

6.5.2.1 Design of S27 Using Cut-Set Retiming

It is also interesting to see the effect when the cut-set retiming design procedure is applied to a nonsystolic architecture in QCA. The cut-set can be chosen such that the subgraph is a single node and the remaining parts form another subgraph. Therefore, in this case, cut-set retiming consists of choosing a node as a cut-set, subtracting one delay from each outgoing edge of the node, and adding one delay to each edge entering the node.

The nonsystolic architecture studied here is the S27 sequential benchmark circuit proposed in ISCAS'89 [15]. The ISCAS'89 sequential benchmark circuits are a set of 31 digital sequential circuits described at the gate level. They were proposed to serve as benchmarks in sequential test generation and scan-based test generation. In the set of ISCAS'89 benchmarks, the letter *S* signifies that the circuit is sequential; the number that follows represents the number of interconnect lines among the circuit primitives.

The CMOS DFG representation of S27 is shown in Figure 6.16. The cut-set retiming design procedure is applied to S27, and the result is compared with previous work [5], in which a delay-matching design method was utilized. The cut-set retiming design procedure for S27 is outlined as follows:

- Step 1: The CMOS DFG of S27 is mapped into a QCA DFG as shown in Figure 6.17

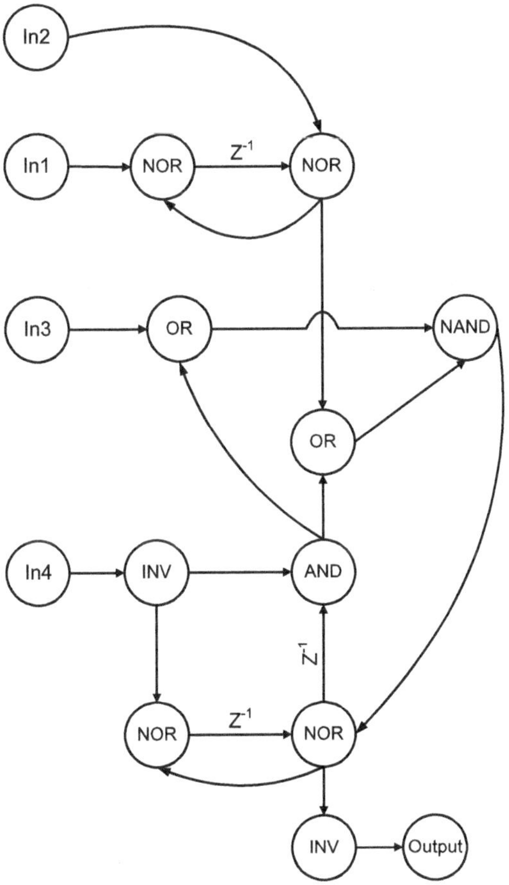

Figure 6.16 CMOS DFG of S27 benchmark.

- Step 2: The logic in the S27 circuit can be optimized for realization with majority logic. As shown in Figure 6.17, the combinational logic that falls within the dashed rectangle can be expressed as follows:

$$\overline{(b+c)(a+c)} = \overline{ab+bc+ac+c} = \overline{M(a,b,c)+c} \tag{6.3}$$

  This optimized logic reduces the combinational circuit complexity from three majority gates to two majority gates. Hence, timing constraints are mitigated.

  However, to offer a fair comparison with the design by Huang [5], the same primitives are used without any logic optimization.

- Step 3: The basic logic units in S27 are inverters (INV), OR gates, AND gates, NOR gates and NAND gates. The NOR (or NAND) gates can

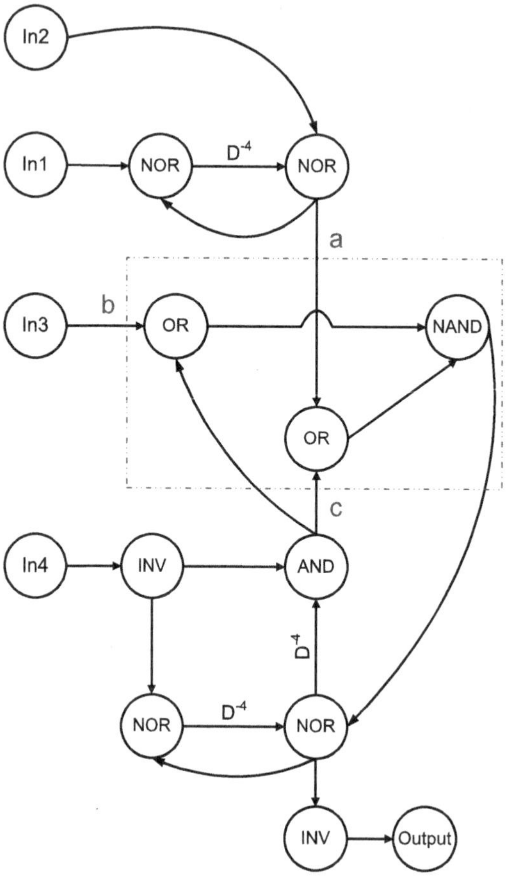

Figure 6.17 QCA DFG of S27 benchmark. (© 2011 IEEE. *From* [2].)

be implemented by combining OR (or AND) gates with inverters. As the design by Huang [5] uses two clocking zone delays (D^{-2}) for each majority gate and one clocking zone delay (D^{-1}) for each inverter, the same timing is used here: D^{-1} for INV, D^{-2} for OR and AND gates, and D^{-3} for NOR and NAND gates.

- Step 4: The worst case loop in the architecture is shown in Figure 6.18. In the loop, the existing delay (E_D) is $4D^{-1}$. To calculate R_D, consider R_{D1} first. There are four gates in the loop, which includes one NOR gate, one AND gate, one OR gate and one NAND gate. According to step 2, these gates require 3 + 2 + 2 + 3 =10 clocking zone delays, (i.e., R_{D1} = 10 D^{-1}). Second, considering an estimate for R_{D2}, the feedback wire from NAND to NOR travels over one OR gate and one AND gate. As the length of each majority gate is five cells as shown in Figure 2.15,

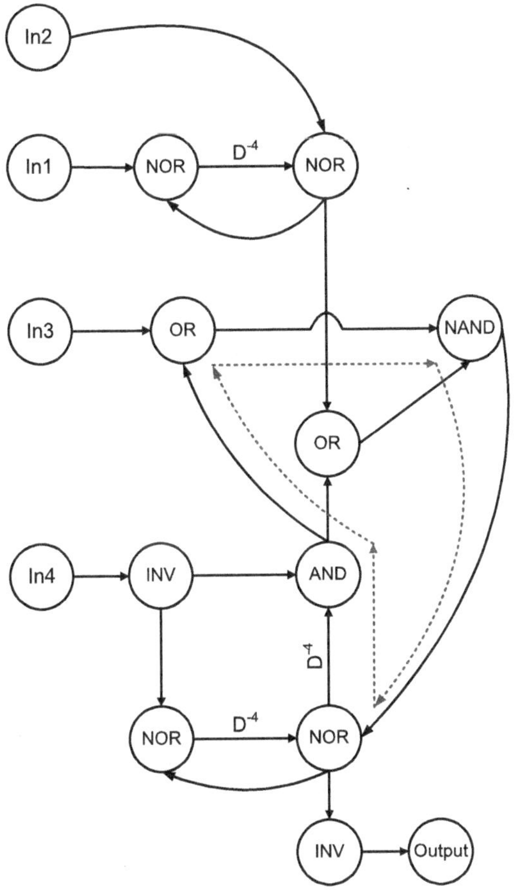

Figure 6.18 Timing analysis: the worst-case loop in S27 architecture. (© 2011 IEEE. *From* [2].)

the feedback wire will travel over a length of $2 \times 5 = 10$ cells, which does not exceed the limitation of 25. As a result, the feedback wire does not require extra delays ($R_{D2} = 0$). The scaling factor is calculated as follows:

$$\alpha = \frac{R_D}{E_D} = \frac{10 + R_{D2}}{4}$$

$$R_{D2} = 0, \text{ therefore, } \alpha = 2.5$$

Thus, all the existing delays in Figure 6.17 should be scaled up by $\lceil \alpha \rceil = \lceil 2.5 \rceil = 3$ times.

- Step 5: 10 clocking zone delays (10 D^{-1}) can be inserted on the forward edges of the four inputs. Next, scaled existing delays are transferred to every logic gate according to its timing requirement, as shown from Figure 6.19 to Figure 6.21. The fully retimed S27architecture is shown in Figure 6.21.

- Step 6: The retimed architecture of S27 can be mapped to a QCA layout as shown in Figure 6.22. It is clear that no wire is longer than 25 cells in this layout. It achieves a constant throughput as long as the pipeline is kept full.
- Step 7: The layout of the QCA S27 architecture is verified using QCA-Designer.

6.5.2.2 Comparison With the Delay-Matching Method

Previous research involved a design of the S27 circuit using a delay-matching method [5]. This method uses conventional flip-flops to ensure that all

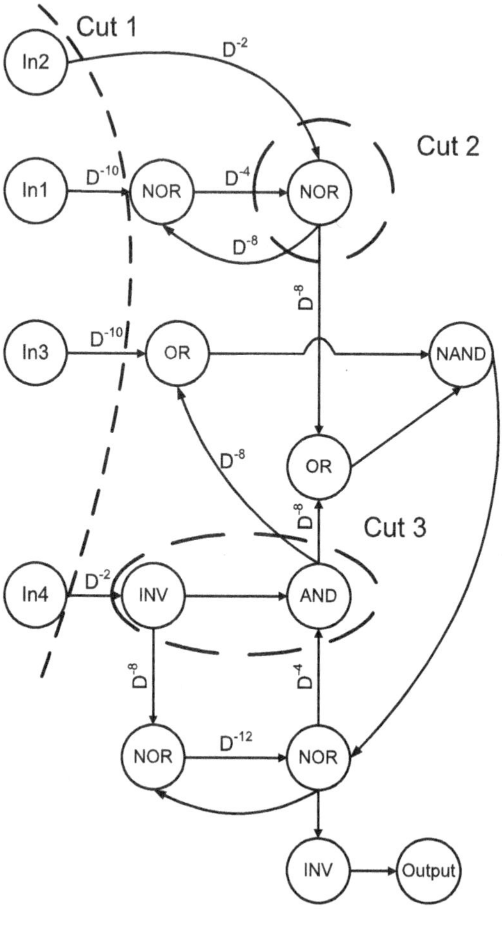

Figure 6.19 Cut-set retiming design procedure of S27 benchmark: cut 1 to cut 3. (© 2011 IEEE. *From* [2].)

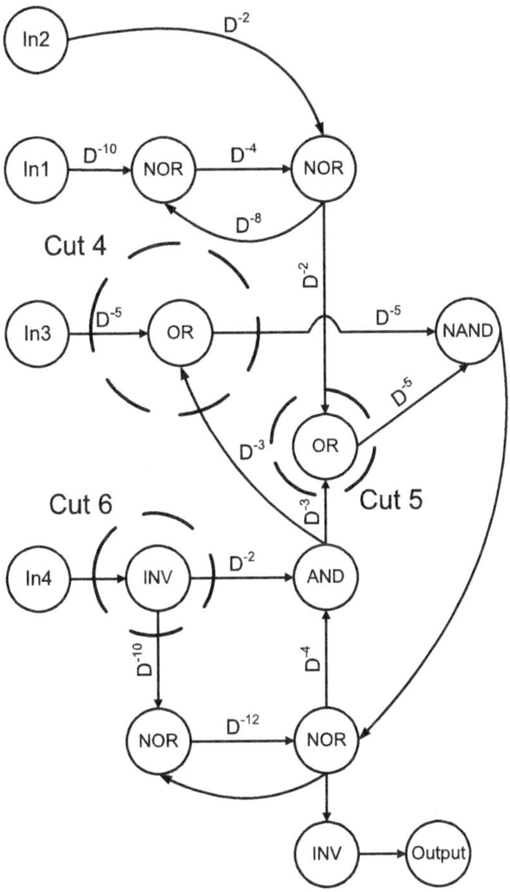

Figure 6.20 Cut-set retiming design procedure of S27 benchmark: cut 4 to cut 6. (© 2011 IEEE. *From* [2].)

the inputs arrive at the QCA flip-flops at the same time. The delay paths are stretched to match the timing requirements. The S27 circuit design using the delay-matching method and the cut-set retiming method are compared using the same primitives (majority gates and inverters) and the same coplanar crossover technique. The comparison is shown in Table 6.1. Note that, for the area comparison, the QCA cell size used in both designs is 18 nm with a center-to-center distance of 20 nm, which is the default setting in QCADesigner.

The value for latency used here is the longest data path from an input to the output without including the loop (from *In*4 to the output). It can be seen from the table that a much more efficient design is achieved using the cut-set retiming design procedure. The cut-set retiming method has a reduction of 22%, 44%, and 46% in terms of cell count, area, and latency of the design, respectively, over the delay-matched design.

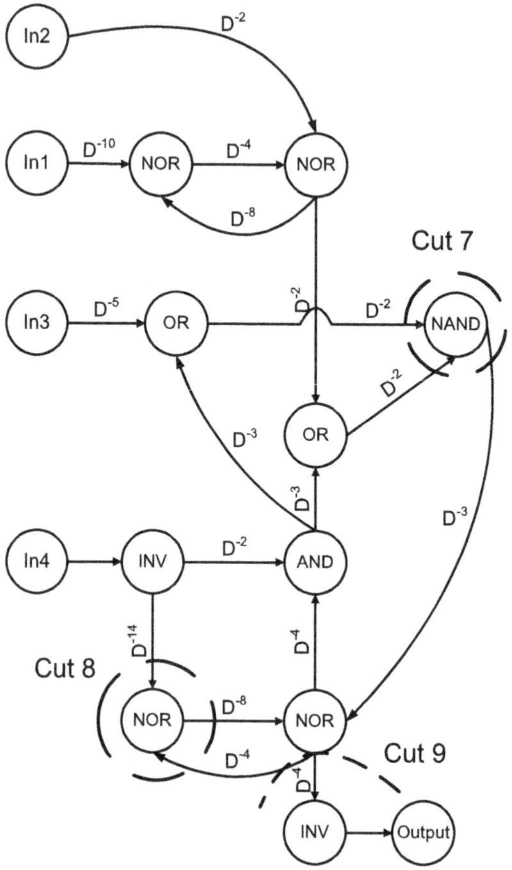

Figure 6.21 Cut-set retiming design procedure of S27 benchmark: cut 7 to cut 9. (© 2011 IEEE. *From* [2].)

In the delay-matched method QCA wires are inserted to achieve even delays, which adds many unnecessary delays, while the proposed cut-set retiming design procedure fully utilizes the existing delays in the original architecture, which minimizes the overall delay, resulting in the reduction in overall complexity and area.

6.6 Conclusion

QCA is a promising alternative to constructing circuits with transistors. However, the unique clocking scheme and wire-level pipelines in QCA present serious timing issues. Consequently, QCA circuit designs encounter timing challenges that make it difficult to assign proper clocking zones and design feedback. This work has examined and discussed the timing constraints and issues associated

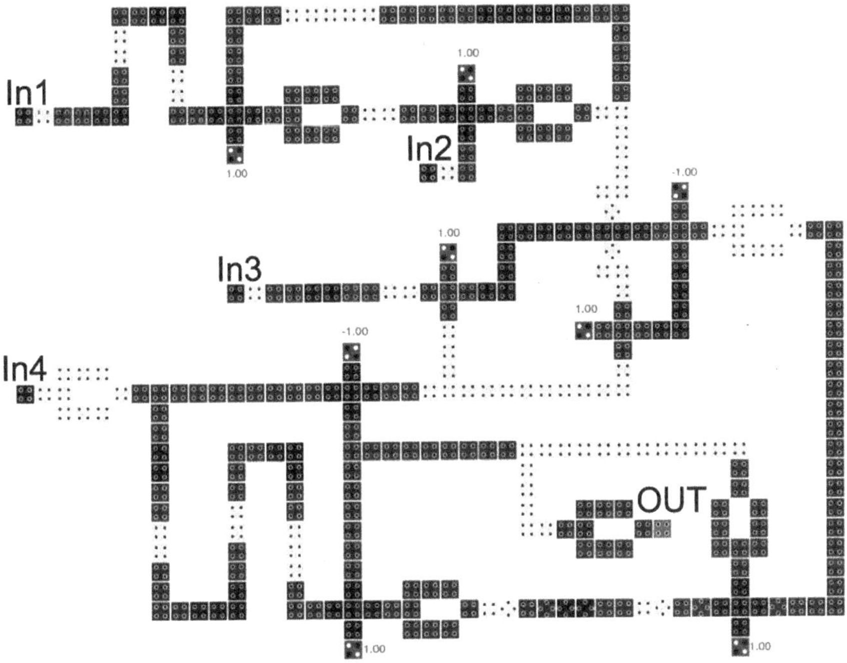

Figure 6.22 QCA layout of the S27 benchmark circuit. (© 2011 IEEE. *From* [2].)

Table 6.1
Comparison Between Delay-Matching Design and Cut-Set Retiming Design

Compared Items	Delay-Matching Design	Cut-Set Retiming Design	Improvement
Complexity	440 cells	340 cells	22%
Area	1.03 μm^2	0.57 μm^2	44%
Latency	12.25 cycles	6.5 cycles	46%

with QCA technology. A cut-set retiming design procedure resolves the timing problems and provides an efficient method to assign the clocking zones that is proposed. Meanwhile, the feedback limitation in QCA is also resolved. Furthermore, this design procedure provides a systematic approach to designing relatively large QCA circuits with complicated timing constraints.

A systolic architecture of a MMM is designed to verify the effectiveness of the proposed design procedure and show that complex cryptographic circuits can be designed in QCA. A nonsystolic S27 benchmark is designed and compared with a previous design. The cut-set retiming method achieves a much more efficient design and provides a reduction of over 20%, 40%, and 40% in terms of the cell count, area, and latency, respectively. The throughput of the QCA S27 architecture is greatly increased by cut-set retiming. From the design results of the case studies, it is clear that cut-set retiming is a very useful method for designing QCA circuits.

References

[1] Liu, W., et al., "Montgomery Modular Multiplier Design in Quantum-dot Cellular Automata using Cut-Set Retiming," in *Proceedings of the 10th IEEE Conference on Nanotechnology,* 2010, pp. 205–210.

[2] Liu, W., et al., "Design of Quantum-dot Cellular Automata Circuits Using Cut-Set Retiming," *IEEE Transactions on Nanotechnology,* Vol. 10, 2011, pp. 1150–1160.

[3] Niemier, M., and P. Kogge, "Problems in Designing with QCAs: Layout= Timing," *International Journal of Circuit Theory and Applications,* Vol. 29, No. 1, 2001, pp. 49–62.

[4] Vankamamidi, V., M. Ottavi, and F. Lombardi, "Two-Dimensional Schemes for Clocking/Timing of QCA Circuits," *IEEE Transactions on Computer-Aided Design of Integrated Circuits and Systems,* Vol. 27, No. 1, 2008, pp. 34–44.

[5] Huang, J., M. Momenzadeh, and F. Lombardi, "Design of Sequential Circuits by Quantum-Dot Cellular Automata," *Microelectronics Journal,* Vol. 38, No. 4-5, 2007, pp. 525–537.

[6] Kung, S., "On Supercomputing with Systolic/Wavefront Array Processors," *Proceedings of the IEEE,* Vol. 72, 1984, pp. 867–884.

[7] Lent, C., P. Tougaw, and W. Porod, "Quantum Cellular Automata: the Physics of Computing with Arrays of Quantum Dot Molecules," in *Proceedings of Workshop on Physics and Computation,* 1994, pp. 5–13.

[8] Walus, K., et al., "Simple 4-bit Processor Based on Quantum-dot Cellular Automata (QCA)," in *Proceedings of the 16th IEEE International Conference on Application-Specific Systems, Architecture Processors,* 2005, pp. 288–293.

[9] Walus, K., and G. Jullien, "Design Tools for An Emerging SoC Technology: Quantum-Dot Cellular Automata," *Proceedings of the IEEE,* Vol. 94, 2006, pp. 1225–1244.

[10] Parhi, K., *VLSI Digital Signal Processing Systems: Design and Implementation,* New York: Wiley, 1999.

[11] Leiserson, C., and J. Saxe, "Retiming Synchronous Circuitry," *Algorithmica,* Vol. 6, No. 1, 1991, pp. 5–35.

[12] Yi, Y., and R. Woods, "Hierarchical Synthesis of Complex DSP Functions Using IRIS," *IEEE Transactions on Computer-Aided Design of Integrated Circuits and Systems,* Vol. 25, 2006, pp. 806–820.

[13] Zhang, R., et al., "A Method of Majority Logic Reduction for Quantum Cellular Automata," *IEEE Transactions on Nanotechnology,* Vol. 3, 2004, pp. 443–450.

[14] Walus, K., et al., "QCADesigner: A Rapid Design and Simulation Tool for Quantum-dot Cellular Automata," *IEEE Transactions on Nanotechnology,* Vol. 3, 2004, pp. 26–31.

[15] Brglez, F., D. Bryan, and K. Kozminski, "Combinational Profiles of Sequential Benchmark Circuits," in *Proceedings of the IEEE International Symposium on Circuits and Systems*, 1989, pp. 1929–1934.

[16] Rivest, R., A. Shamir, and L. Adleman, "A Method for Obtaining Digital Signatures and Public-Key Cryptosystems," *Communications of the ACM*, Vol. 21, No. 2, 1978, pp. 120–126.

[17] Miller, V., "Use of Elliptic Curves in Cryptography," in *Proceedings of Advances in Cryptology*, 1986, pp. 417–426.

[18] Montgomery, P., "Modular Multiplication without Trial Division," *Mathematics of Computation*, Vol. 44, No. 170, 1985, pp. 519–521.

[19] Marnane, W., "Optimised Bit Serial Modular Multiplier for Implementation on Field Programmable Gate Arrays," *Electronics Letters*, Vol. 34, No. 8, 1998, pp. 738–739.

[20] Cho, H., and E. Swartzlander, Jr., "Adder and Multiplier Design in Quantum-dot Cellular Automata," *IEEE Transactions on Computers*, Vol. 58, 2009, pp. 721–727.

7

QCA Systolic Array Design[1]

Weiqiang Liu, Liang Lu, Máire O'Neill, and Earl E. Swartzlander, Jr.

7.1 Introduction

As an emerging post-CMOS technology, it is interesting to explore the characteristics of QCA. Previous research [4–6] has shown that in the case of an adder design, the propagation delay is strongly dependent on the density of the QCA layout. As discussed in Chapter 3, in traditional CMOS technology, CLA adders are faster than RCAs. However, in QCA a carefully optimized layout of a 64-bit RCA is 50% faster than a 64-bit CLA adder [5]. The wire delays in QCA account for most of the difference. Since an adder is a relatively simple component in digital circuit design, it is envisaged that in the case of a complex design, wire delay, which results in extra clock cycles in QCA technology, will seriously affect the performance of a QCA system. Therefore, this chapter investigates more complex architectures to further explore the characteristics of QCA technology.

To improve the performance of a computing system, a common solution is to employ a systolic array architecture. Based on the data flows of systolic arrays, two types of systolic array architecture are studied in this chapter. In the first type, input data flows through the array on each clock cycle and the results remain in the processing elements (PEs). A typical example is a systolic matrix multiplier. In the second type, the results flow from one PE to the next PE. In other words, results of PEs flow through the entire array. An example application of this type is a Galois field multiplier. The main difference between these two types of systolic arrays is that if they are modified into a single processor architecture, the first one can also be pipelined. However, the second one

1. Chapter 7 is based on [1–3].

requires global feedback to achieve its functionality, which involves much more complicated control logic. These two cases will be discussed separately with two case studies, concerning a matrix multiplier and a Galois field multiplier. Comparisons between coplanar and multilayer crossing based designs are provided to show the impact of crossover options on QCA design results.

This chapter is organized as follows: Section 7.2 introduces signal-flow graphs (SFGs) and defines systolic array architectures. Sections 7.3 and 7.4 present case I, the matrix multiplier, and case II, the Galois field multiplier, along with their implementations and analysis, respectively. Section 7.5 provides a general methodology for QCA systolic array design. Finally, Section 7.6 presents conclusions.

7.2 Signal Flow Graph and Systolic Array Architecture

7.2.1 Signal Flow Graph

The SFG with a collection of nodes and directed edges is another graphic representation approach for algorithms [7]. The nodes represent computations and the directed edges denote linear transformations from nodes to nodes. SFGs are used for analysis, representation and evaluation of linear digital networks, in which the edges are usually restricted to constant gain multipliers or delay elements. Adders and multipliers can be described by a node with multiple incoming edges and one outgoing edge. For the same example used in the introduction of the DFG, the SFG of the computation $y(n) = ay(n-1) + x(n)$ [corresponding to the block diagram in Figure 7.1(a)] is shown in Figure 7.1(b), in which an edge with no explicit indication of operation represents an edge with unit-gain (i.e., identity transformation).

7.2.2 Systolic Array Architecture

Systolic array architectures, were first proposed in 1978 by Kung and Leiserson [8, 7]. The name derives from an analogy with the regular pumping of blood by the heart. Systolic array architectures represent a network of PEs that rhythmically compute and pass data through the system. Data flows synchronously across the array between neighbors, usually with different data flowing in different directions. Each processor takes input data from neighbors, processes them, and outputs the results. Processors in the array compute data and store them independently of each other. In this pipelined manner, systolic arrays can be extremely fast and are easily scalable in comparison to single processor machines. Due to this pipelined manner, data flow control becomes much simpler. Fewer branches are required in such systolic array architectures; therefore, the cost in control signal generation or finite state machine (FSM) design is

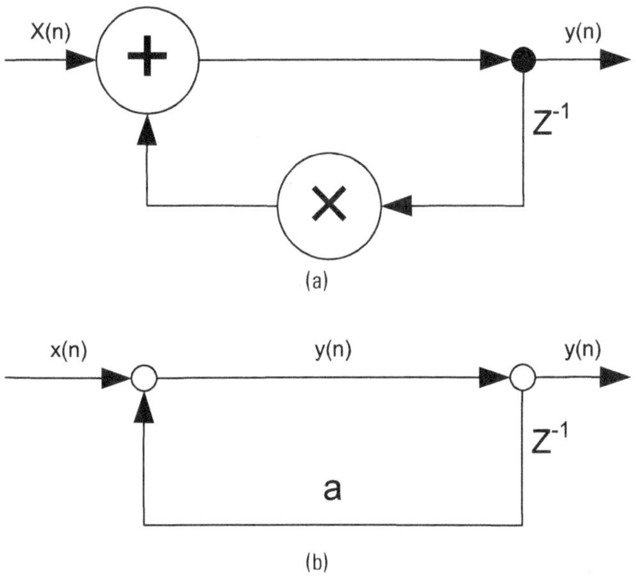

Figure 7.1 Computation $y(n) = ay(n-1) + x(n)$: (a) block diagram, and (b) SFG representation.

significantly reduced. The main design effort of systolic array architectures is mapping [7]. Although they are application specific, systolic arrays are widely used in cryptographic algorithms, signal processing, and other computationally intensive tasks.

Based on the number of PEs, architectures are divided into two categories in this chapter: systolic array architectures which consist of more than one PE, and nonsystolic array architectures which have only one PE and are thus referred to as single processor architectures.

7.3 Case Study I: Matrix Multiplier

7.3.1 Systolic Array Matrix Multiplier Introduction

Matrix computations are widely used in information processing. Let $A = (a_{ik})$ and $B = (a_{kj})$ be two rectangular matrices of order $N_1 \times N_3$ and $N_3 \times N_2$, respectively. Their product, matrix $C = A \times B$, $C = (c_{ij})$, can be obtained according to the following recurrence relationship:

$$\begin{aligned} &c_{ij}^{(k)} = c_{ij}^{(k-1)} + a_{ik} \cdot b_{kj} \ \textit{where} \ c_{ij}^{(0)} \equiv 0,\ c_{ij} \equiv c_{ij}^{(k)} \\ &i = 1,2,\ldots,N_1,\ j = 1,2,\ldots,N_2 \text{ and } k = 1,2,\ldots,N_3 \end{aligned} \tag{7.1}$$

This algorithm is easily mapped to a systolic array architecture [8], as depicted in Figure 7.2, which comprises a matrix of processing elements. Input data are pumped into this array and pass across the array. These data are calculated in isolation in each PE. The final value of each element in the resulting matrix will be generated from each PE. One clock cycle delay is required for input data storage. In the case of a 2 × 2 matrix, four clock cycles are needed to perform the computation. As is well-known, systolic arrays have parallel structures with pipelined data flows. Hence, a case study of the QCA design of a systolic matrix multiplier is analyzed in the following sections.

7.3.2 QCA Systolic Matrix Multiplier Design

In this section, a systolic matrix multiplier based on QCA technology is described. Due to the complexity of QCA circuit design, this systolic array exploration begins with a small matrix size with small bit-width elements. In this case, a 2 × 2 matrix with 2-bit elements is employed.

The matrix multiplier PE architecture is shown in Figure 7.3. Each PE consists of a 2-bit multiplier and an accumulator. The output of the multiplier is 4 bits wide. Since the matrix size is only 2 × 2, only one addition is required during the whole process. Therefore, the length of feedback is 4 bits and the output length of the accumulator is 5 bits. One 2-to-1 MUX is used to control the input of the accumulator.

7.3.2.1 Serial-Parallel Multiplier

The serial-parallel multiplier proposed in [6] is used for multiplication. For N-bit inputs, the multiplier receives $N+1$ inputs and produces a serial output. Both the serial input and output are ordered from LSB to MSB and parallel inputs are repeated whenever a new serial input is provided. This multiplier

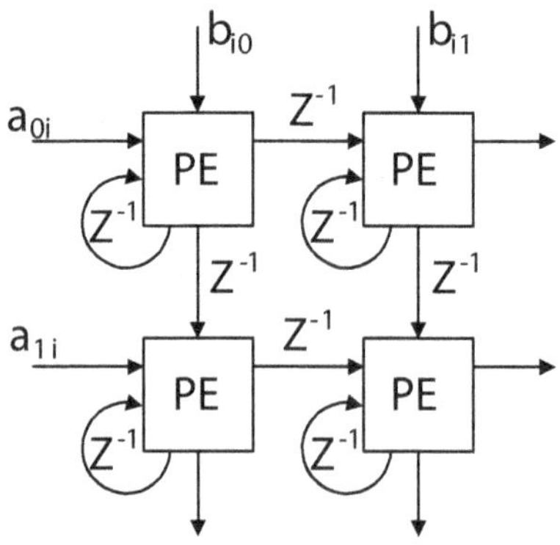

Figure 7.2 Systolic array example: signal flow graph of 2 × 2 matrix multiplier. (© 2013 IEEE. *From* [3].)

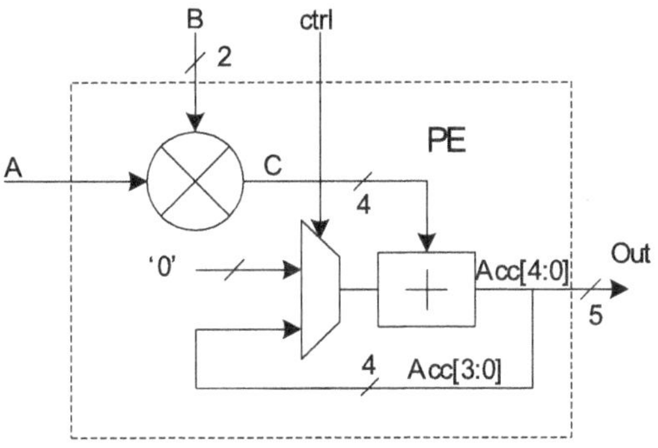

Figure 7.3 Systolic array PE. (© 2013 IEEE. *From* [3].)

can be fully pipelined and is easily scalable. In the systolic PE design, such a 2-bit serial-parallel multiplier is employed. The schematic and layouts based on multilayer crossovers and coplanar crossings are depicted in Figures 7.4–7.6, respectively. In the layout diagrams, four clocking zone phases are indicated using different shades of color. In Figure 7.5, the solid cells indicate those cells in the crossover layer. In addition, the input and output cells are indicated by A, B_0, B_1, and serial out.

When designing a QCA circuit with coplanar crossings, it is essential to carefully arrange the clocking zones on both sides of the crossings. To achieve a robust coplanar crossing, it is suggested to have different clocking zones at the crossing point for the regular cells and the rotated cells. This prevents these two types of cells from interfering with each other. For example, in Figures 7.4– 7.6, four crossings are numbered 1 to 4. At crossing 1, in the multilayer crossover layout, both the vertical and horizontal wires are assigned to clocking zone 0.

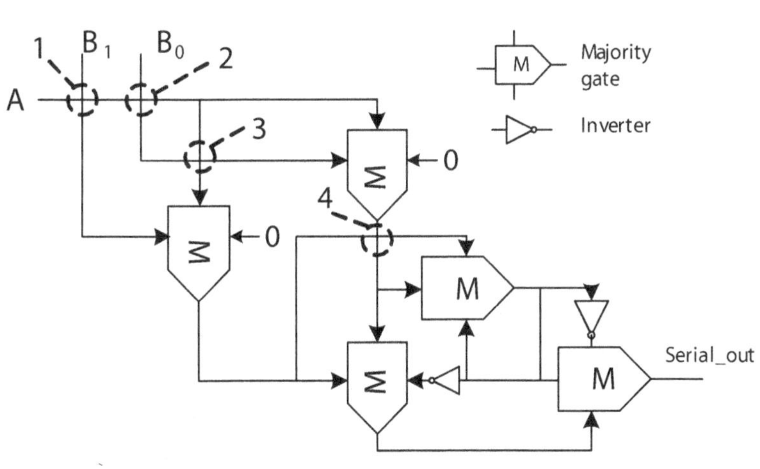

Figure 7.4 2-bit multiplier schematic. (© 2013 IEEE. *From* [3].)

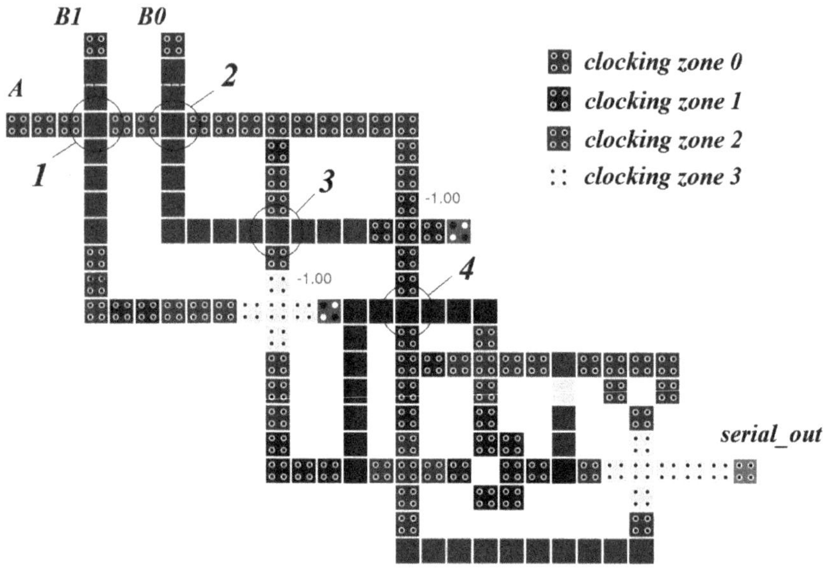

Figure 7.5 2-bit multiplier multilayer crossover layout. (© 2013 IEEE. *From* [3].)

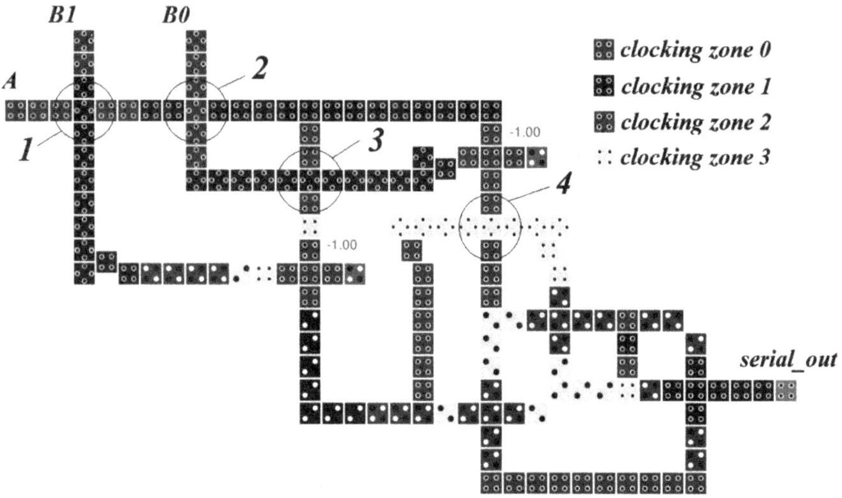

Figure 7.6 2-bit multiplier coplanar crossing layout. (© 2013 IEEE. *From* [3].)

When converted to the coplanar crossing layout, one clocking zone delay is added to the rotated cells (the horizontal wire in Figure 7.6) to ensure that this crossing drives the signals properly. Similarly, at crossing 2, one clocking zone delay is added in the regular cells in the coplanar crossing design. At crossings 3 and 4, both directions have different clocking zones in the multilayer crossover design. Therefore, no extra delay is required. Note that sometimes, extra delays inserted in earlier crossings could affect the clocking zone distribution of later crossings on the data path. Therefore, a careful analysis of clocking zone delays is required when designing circuits using coplanar crossings. When comparing these two layouts of the 2-bit multiplier, the coplanar crossing design has two more clocking zone delays than the multilayer crossover design. In addition, due to the different clocking zone assignments needed in the coplanar crossing design, extra spaces are required to ensure that the circuit operates correctly. For instance, at the top left area of the layout, the distance between the two vertical input wires is increased from 2 cells to 4 cells. Therefore, in general, a coplanar crossing design tends to have a larger area than a multilayer crossing design.

7.3.2.2 Accumulator

The accumulator design is based on a carry flow adder (CFA) [6]. A MUX is required to choose the correct input to the accumulator. Its input is either "0" or a previous sum. Therefore, the MUX architecture can be simplified to a majority gate in QCA. Two inputs are the signals to be chosen and the third input is the selector. If the selector is set to "0," then the output of this majority gate will be "0" no matter what the value of the other input. If the selection signal is set to "1," then the output will be the previous sum. For the purpose of explanation, instead of using a 4-bit accumulator, Figures. 7.7 and 7.8 show a schematic and a QCA multilayer crossover layout for a 2-bit accumulator, respectively. It has also been designed using the coplanar crossing scheme. Nine majority gates and four inverters are used in this accumulator—namely, M_1 to M_9 and I_1 to I_4. M_2, M_3, M_4, I_1, I_2 and M_7, M_8, M_9, I_3, I_4 are used to construct the LSB and

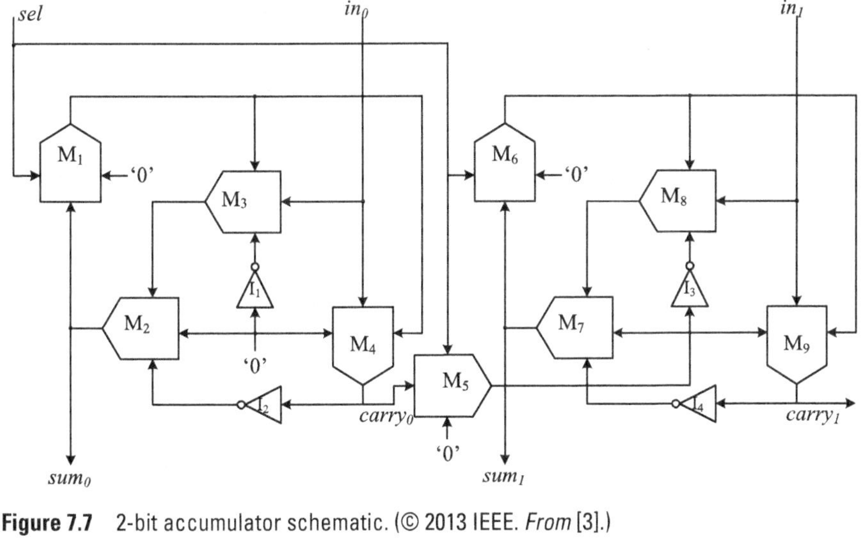

Figure 7.7 2-bit accumulator schematic. (© 2013 IEEE. *From* [3].)

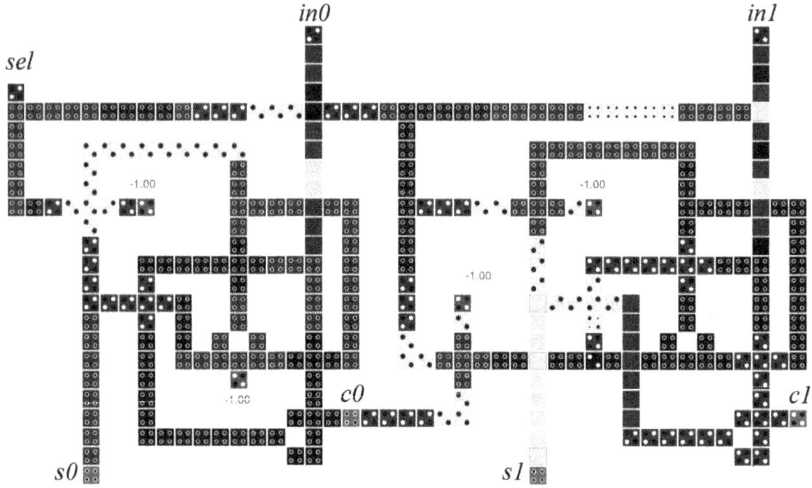

Figure 7.8 2-bit multilayer accumulator layout. (© 2013 IEEE. *From* [3].)

MSB of the CFA adder, respectively. One input of the CFA is in_0/in_1, while the second input is chosen by M_1/M_6. For example, at the initial stage, *sel* is set to "0"; therefore the second input will be "0." During the accumulating process, feedback sums come from M_2/M_7 and will be selected as the second input to be added with the new input in_0/in_1. M_5 is used to control the carry from the LSB adder. It is manipulated by sel in the same manner as M_1 and M_6. Since the wires at all crossings have different clocking zones in the multilayer design, no extra clocking zones are required for the coplanar design.

7.3.2.3 Serial-to-Parallel Converter

A serial-to-parallel converter is required in the design of the systolic matrix multiplier architecture. Apart from the *sel* signal used within the accumulator, another control signal, *ctrl*, is employed to guarantee a correct data flow. The schematic and multilayer crossover layout of this part is shown in Figures 7.9 and 7.10. Due to the characteristics of the bit-serial multiplier, the serial output has to be synchronized first. Therefore, a delay wire is applied at the output of the multiplier which functions as a serial-to-parallel converter. As shown in the layout, each bit of the bit-serial multiplier output has a quarter-cycle difference. This pipelined delay line introduces another issue: *p*3 will be input into *p*0, for instance, when it is available from the bit-serial multiplier output. Hence, a simple control signal is required to select the correct input for the accumulator. A design approach similar to that used in the accumulator is employed. A majority gate set to be an AND gate to compute p and the ctrl signal ensures that the correct data flow is maintained to the accumulator. As illustrated in the layout, each bit includes such a majority gate. Once a corrected result arrives,

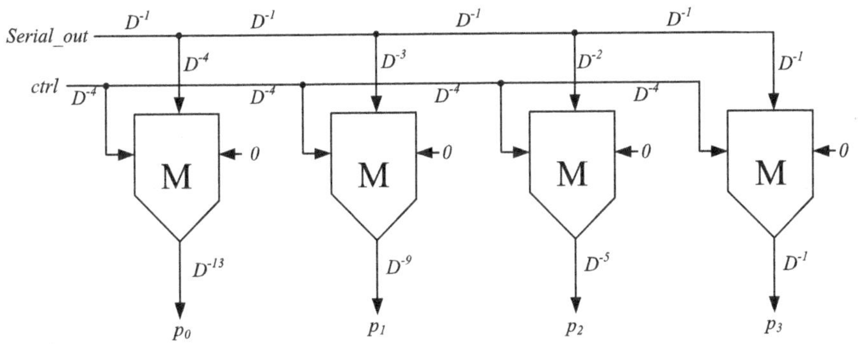

Figure 7.9 Serial-to-parallel converter schematic. (© 2013 IEEE. *From* [3].)

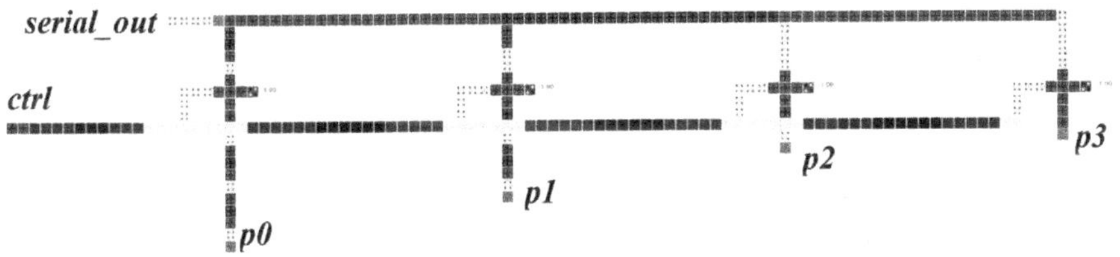

Figure 7.10 Serial-to-parallel converter multilayer layout. (© 2013 IEEE. *From* [3].)

the majority gate will allow it to continue through, otherwise, "0" is selected as the input for the accumulator. Again, in the coplanar design of this module, no extra clocking zones are required.

7.3.2.4 2-Bit 2 × 2 Matrix Multiplier Systolic Array Design

The QCA multilayer layouts of one PE and the whole systolic 2-bit 2 × 2 matrix multiplier are shown in Figures 7.11 and 7.12. Four PEs are placed in a square array. Each PE includes a 2-bit bit-serial multiplier (top left of each PE) and a 4-bit accumulator. In between, a serial-to-parallel converter is used to synchronize the serial output and correctly propagate the corresponding results. Input signals are loaded into the systolic array from the top left of this circuit. The horizontal input is the serial input A and the vertical input is the 2-bit parallel input B. Between each PE, a one clock cycle delay is set for the systolic array in both the vertical and horizontal directions. To output the result in parallel, delays are required at the output. The *ctrl* and *sel* signals are set for each PE to guarantee a correct data flow.

This architecture can be easily scaled in terms of matrix size and data word size. To explore the characteristics of the QCA systolic array, various matrix sizes and word sizes have been designed and simulated in both multilayer and coplanar designs. Since the current version of QCADesigner [9] has difficulties in designing and simulating very large circuits, the following matrix sizes were investigated:

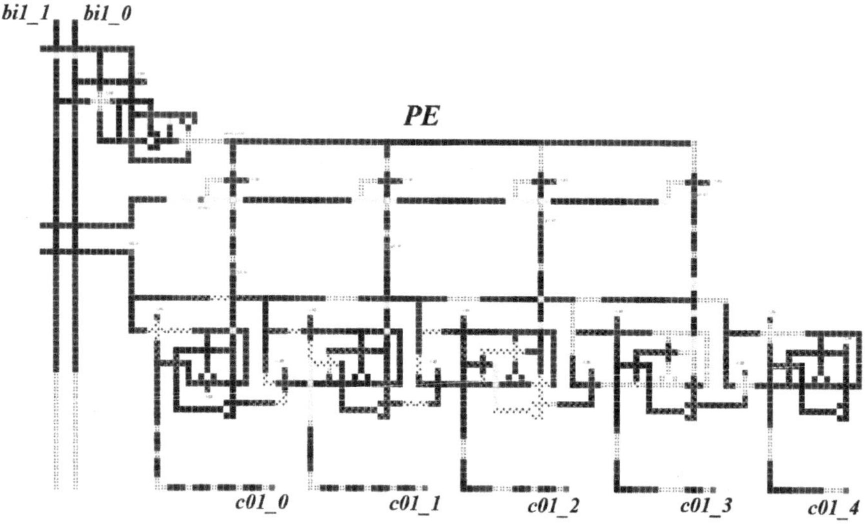

Figure 7.11 One PE multilayer layout of the 2-bit 2 × 2 systolic matrix multiplier.

{2 × 2, 3 × 3, 4 × 4} matrices with 2-bit elements and a 2 × 2 matrix with data word sizes of {2-bit, 3-bit, 4-bit}. The comparison and analysis of these circuits are discussed in Section 7.3.3.

7.3.3 Design Study

7.3.3.1 Simulation and Results

The matrix multipliers of the previous section were simulated using QCADesigner. In the simulations, most default parameters for a bistable approximation as listed in Section 2.4.1.3 are used. The number of samples is determined to be 800,000.

In the simulation, it takes 20 clock cycles to finish the multilayer crossing design of the 2-bit 2 × 2 matrix multiplication. In the case of the coplanar design, 20.5 clock cycles (0.25 of a clock cycle indicates one clocking zone delay) are required to complete this matrix multiplication. These designs have been extended in terms of word size and matrix size.

The design results are shown in Table 7.1. In the coplanar crossing designs, as discussed above, the side margin spacing near the crossings needs to be larger (for example, the spacing between the vertical input wires in Figure 7.6). These larger spacings cause the design to be somewhat bigger and slightly slower. In comparison to the coplanar crossing designs, the cell counts and areas of the multilayer crossover designs are reduced by approximately 4.1% and 17.4% on average, respectively. The delay is also reduced by 2% to 3% in the multilayer crossover designs.

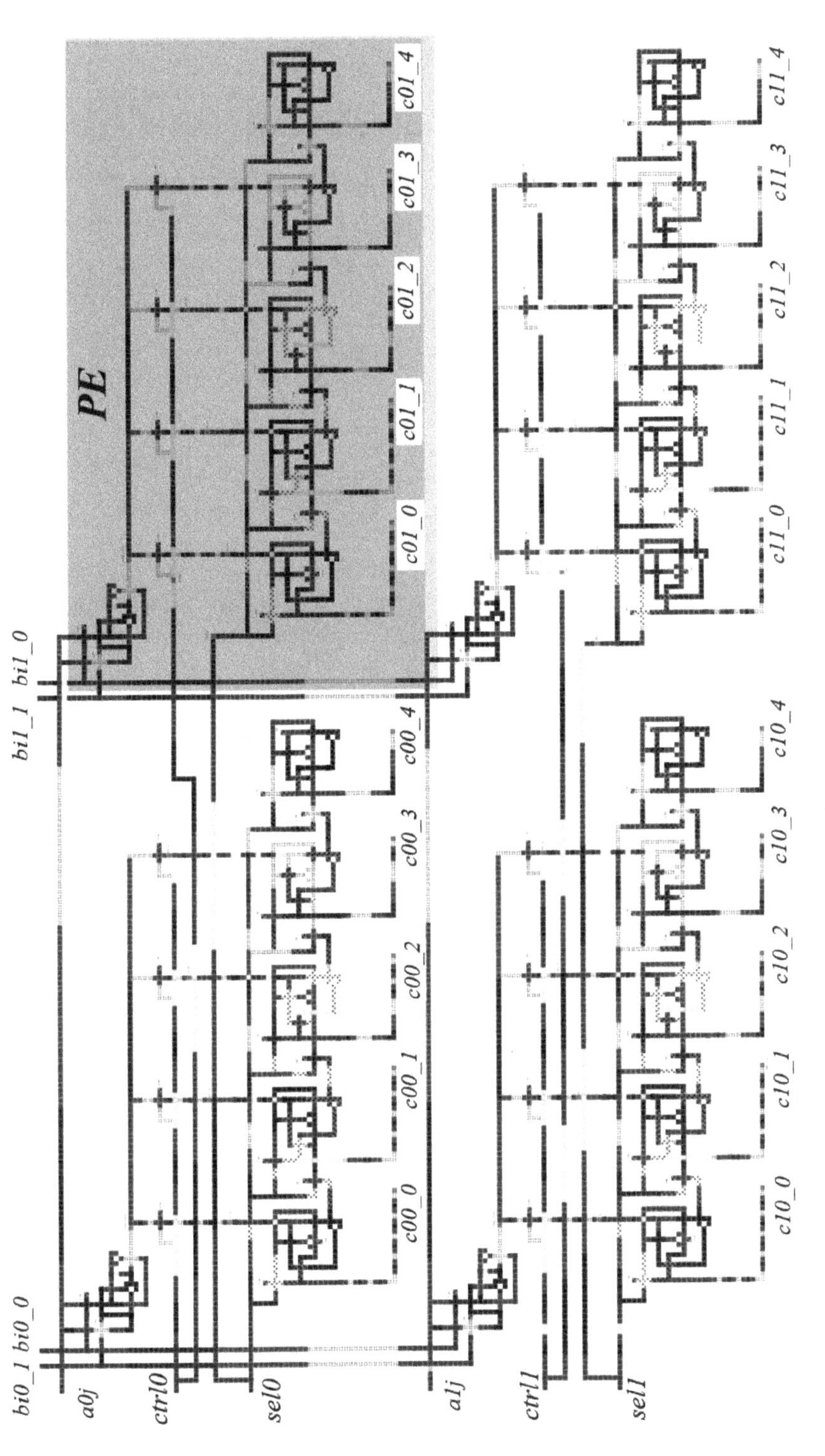

Figure 7.12 Multilayer layout of the 2-bit 2 × 2 systolic matrix multiplier. (© 2013 IEEE. *From* [3].)

Table 7.1
Matrix Multiplier With Different Word Sizes/Matrix Sizes

	Designs	Complexity (Number of Cells)	Area (μm^2)	Delay (Number of Cycles)
Multilayer	2-bit 2 × 2	7,102	15.69	20
	3-bit 2 × 2	11,361	25.05	26
	4-bit 2 × 2	15,759	38.96	32
Coplanar	2-bit 2 × 2	7,300	19.23	20.5
	3-bit 2 × 2	11,438	29.54	26.75
	4-bit 2 × 2	16,230	45.49	33
Multilayer	2-bit 2 × 2	7,102	15.69	20
	3-bit 2 × 2	16,690	36.51	26
	4-bit 2 × 2	34,199	75.02	33
Coplanar	2-bit 2 × 2	7,300	19.23	20.5
	3-bit 2 × 2	17,359	43.23	26.5
	4-bit 2 × 2	36,025	92.50	33.5

The resources used in the QCA layout of each design are listed in Table 7.2. Note that both designs utilize the same number of majority gates and crossings. However, since extra inverters are required for signal transfer from rotated wires to regular wires, there are different numbers of inverters and wire cells in each design. Many signal transfer inverters are required for the coplanar design. The inverter and wire cell counts in the table for the multilayer and coplanar designs are indicated as MC and CC, respectively.

Although the practicality of a multilayer crossover is still an open issue, it is more stable in terms of error tolerance for random cell displacement. Previous work [10] has examined the possibility of multilayer QCA, but there has been no reported implementation of this crossover. This research shows that there is not much difference in terms of circuit area and delays for different crossover types.

Table 7.2
Matrix Multiplier Implementation Resource Count

Designs	Number of M	Number Inv MC/CC	Number of Crossings	Number of Wire Cells MC/CC
2-bit 2 × 2	132	48/188	124	6,106/6,184
3-bit 2 × 2	196	72/292	192	9,877/9,762
4-bit 2 × 2	260	96/388	270	13,787/14,010
2-bit 3 × 3	267	108/429	304	14,449/14,836
2-bit 4 × 4	528	192/864	569	30,215/31,273

MC=multilayer, CC=coplanar

7.3.3.2 Analysis and Comparison

One purpose of this case study is to explore the QCA characteristics for systolic array design, in which only the input data pass across the PEs. As described in Section 7.1, high-throughput systolic arrays benefit from their pipelined nature and parallel computing strategy. In this section, a comparison is provided to ascertain whether these benefits also apply to QCA technology.

In order to perform a matrix multiplication, a single processor architecture needs to use an iterative data flow to complete one calculation. In contrast, a systolic array design involves multiple PEs operating in parallel. Therefore, compared to a single PE architecture, the systolic array significantly reduces the clock cycle count. It is unfair to directly compare a QCA-based design with a CMOS-based design, since they are very different technologies. Clock cycles are defined differently in each of these technologies. Hence, the ratio of single processor clock cycle count to systolic array clock cycle count (single processor/systolic array) is used as a comparative metric as shown in Table 7.3. A larger (single processor/systolic array) ratio indicates that more speed benefits are achieved by applying a systolic array. The clock cycle counts indicate the number of clock cycles required to complete one matrix multiplication.

The single processor architecture employs iterative processing with a single PE. In this case, the single processor architecture can still be pipelined due to its non-feedback architecture. Therefore, no extra control circuits are needed to manage the data flow. For comparison purposes, CMOS-based matrix multipliers are also implemented. In the case of the CMOS-based architectures, one PE includes a multiplier and an accumulator. After accumulating the products, the result is stored in a register. For example, a 2×2 matrix multiplier calculation has eight multiplications. Therefore, eight clock cycles are required to complete this calculation. In the systolic array case, only four clock cycles are required, giving a clock cycle count ratio of $8/4 = 2$. In the QCA-based architectures, the single processor clock cycle count value refers to the number of clock cycles required in a single PE times the number of PEs. For example, in the case of the

Table 7.3
Clock Cycle Count Ratios of CMOS vs QCA Design

Designs		**CMOS**			**QCA-Multilayer**			**QCA-Coplanar**		
Word Size	**Matrix Size**	**SP**	**SA**	**Ratio (SP/SA)**	**SP**	**SA**	**Ratio (SP/SA)**	**SP**	**SA**	**Ratio (SP/SA)**
2-bit	2×2	8	4	2	72	20	3.60	74	20.5	3.61
3-bit		8	4	2	96	26	3.69	99	26.75	3.70
4-bit		8	4	2	120	32	3.75	124	33	3.76
2-bit	1×1	2	2	1	18	18	1.00	18.5	18.5	1.00
	2×2	8	4	2	72	20	3.60	74	20.5	3.61
	3×3	27	7	3.8	198	26	7.62	202.5	26.5	7.64
	4×4	64	10	6.4	432	33	13.09	440	33.5	13.13

multilayer design, in a 2-bit 2 × 2 matrix multiplier calculation, one PE performs two multiplications and one accumulation in 18 clock cycles. Therefore, to finish the entire matrix, it takes 18 × 4 = 72 cycles. As the corresponding systolic array design takes 20 cycles, the clock cycle count ratio is 72/20 = 3.60. The clock cycle counts listed in Table 7.3 are based on actual designs.

Note that the coplanar and multilayer QCA architectures have almost the same delay. This is because in these designs, coplanar crossings only introduce extra delays in the input wires. One extra clocking zone delay is required as the word size increases by one bit. Compared to the overall delays, these extra delays are negligible in this case. Therefore, in a large architecture, by carefully arranging the clocking zones into small zones it is possible to reduce the impact of the extra delays caused by coplanar crossings.

A comparison of the clock cycle ratio as a function of the word size is shown in Figure 7.13. In the CMOS designs, the number of cycles remains constant as the word size increases. However, in the QCA designs, larger word sizes require a larger accumulator. This results in longer internal wire delay. Therefore, the clock cycle count in the single processor design increases significantly. In the systolic array, because of its inherent parallelism, the wire delay does not magnify as much as for the single processor architecture. This leads to an increase in the clock cycle ratio. Therefore QCA technology benefits from the systolic array architecture as the word size increases. Note that when compared to the multilayer design, the coplanar design only introduces limited clock cycle delay at the input stage in this case, where each extra bit requires

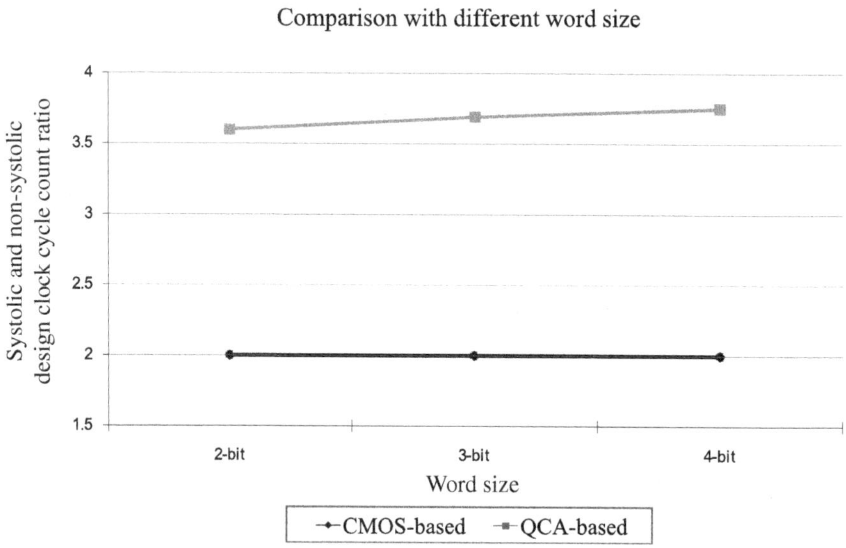

Figure 7.13 Clock cycle count ratio versus word size. (© 2013 IEEE. *From* [3].)

one additional clocking zone delay. This does not have much of an effect on the single processor/systolic array ratio. Hence, the ratio trends of the multilayer and the coplanar QCA designs are effectively the same.

Figure 7.14 shows the clock cycle ratio as a function of the matrix size. It is clear that the systolic array designs benefit from their parallelism and pipelined capacities in both technologies. It is anticipated that the level of parallelism will be determined by the matrix size, since as more PEs are used, the architecture achieves higher parallelism. In the CMOS architecture, the single processor design requires more clock cycles for a larger matrix due to a greater number of additions required in the accumulators, while due to the inherent parallelism, in the systolic array design, the increase in the number of clock cycles is less. Therefore, the clock cycle ratio in the CMOS design increases.

In the QCA architecture, the single PE delay increases because of the increasing size of the accumulator. However, the increase in clock cycle ratio is significantly greater than in the CMOS design. This is because in the CMOS design, one more addition only costs one more clock cycle. However, in the QCA design, internal wire delays cost more clock cycles. This disadvantage of the internal delay is further magnified by larger matrices due to the large number of iterations needed. In the QCA systolic array design, the clock cycle count increases as in the CMOS design, because there is only one clock cycle delay between each PE in both technologies. Therefore, the clock cycle ratio increases much faster in the QCA designs than in the CMOS designs. In other words, QCA technology benefits much more from employing systolic array architectures than CMOS technology.

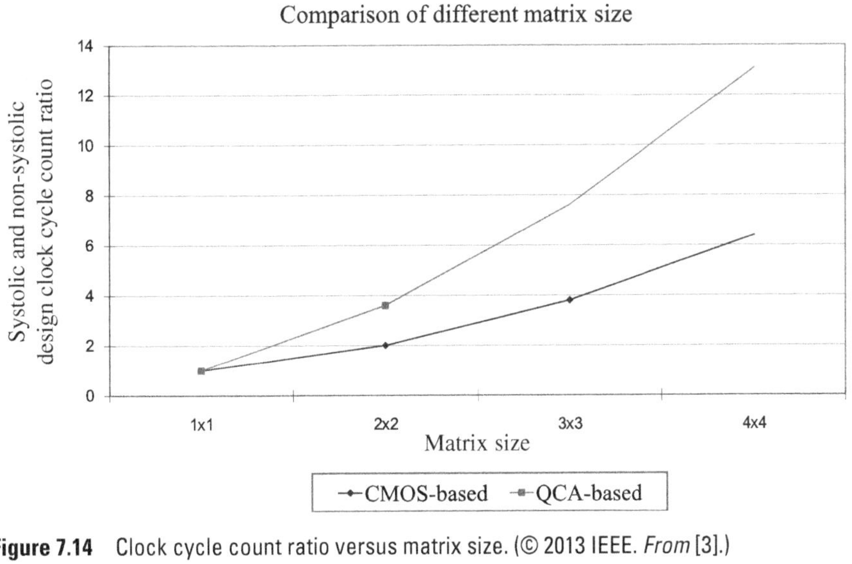

Figure 7.14 Clock cycle count ratio versus matrix size. (© 2013 IEEE. *From* [3].)

Based on these results, it is clear that a systolic architecture can reduce the impact of the wire delay in QCA technology, due to its inherent parallelism and pipelining, especially for large arrays. QCA technology is a terahertz technology and although QCA designs require more clock cycles than the equivalent CMOS architectures, they run much faster [11].

It is interesting to see if the speed benefit of applying systolic array to QCA technology is achieved by using more resources. The ratio of systolic array area to single processor area (systolic array/single processor) is used as a comparative metric to reflect the resources used in each technology as shown in Table 7.4. A larger systolic array/single processor ratio indicates that more resources are used. Both clock cycle count ratio and the area ratio are used to define the efficiency of a design, which is shown as follows:

$$\eta = \frac{C}{A} \tag{7.2}$$

where, η is the efficiency, C is clock cycle count ratio, and A is area ratio. The efficiency of each design in both CMOS and QCA are shown in Table 7.5. A comparison of the efficiency as a function of the word size is shown in Figure 7.15. It can be seen that the efficiencies of both technologies increase significantly; however, the QCA design is much more efficient than its CMOS counterpart as the word size increases. Figure 7.16 shows the efficiencies as a function of matrix size. It is shown that the efficiencies of both technologies generally decrease as the matrix size increases. However, the efficiency of the CMOS design decreases much more rapidly than that of the QCA design. The comparisons show that the QCA systolic array design is generally efficient.

Table 7.4
Area Ratios of CMOS versus QCA Design (in micrometers)

Designs		**CMOS-based (65 nm)**			**QCA-Multilayer (20 nm)**		
Word Size	**Matrix Size**	**SP**	**SA**	**Ratio (SP/SA)**	**SP**	**SA**	**Ratio (SP/SA)**
2-bit	2×2	91.5	574.7	6.28	3.94	15.72	3.99
3-bit		167.7	1008.3	6.01	6.30	24.88	3.95
4-bit		328.3	1679.0	5.11	9.73	38.96	4.00
2-bit	1×1	91.5	91.5	1.00	3.94	3.94	1.00
	2×2	91.5	574.7	6.28	3.94	15.72	3.99
	3×3	91.5	1735.7	18.97	3.94	36.38	9.23
	4×4	108.2	3590.1	33.18	5.21	74.99	14.39

Table 7.5
Efficiency of CMOS versus QCA Design

Designs		**CMOS-based (65 nm)**	**QCA-Multilayer (20 nm)**
Word Size	**Matrix Size**	η	η
2-bit	2×2	0.32	0.90
3-bit		0.33	0.93
4-bit		0.39	0.94
2-bit	1×1	1.00	1.00
	2×2	0.32	0.90
	3×3	0.20	0.83
	4×4	0.19	0.91

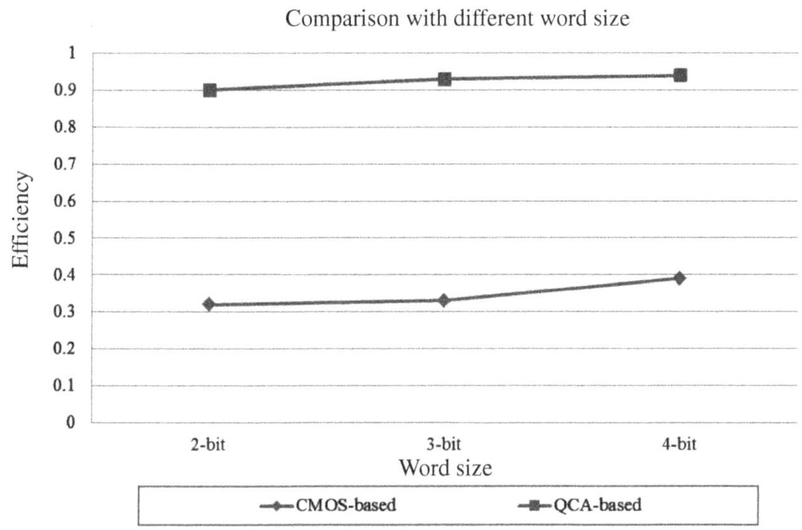

Figure 7.15 Efficiency versus word size.

7.4 Case Study II: Galois Field Multiplier

7.4.1 Galois Field Multiplier Introduction

A Galois field multiplier has wide application in coding theory and cryptography. It consists of a finite set of elements together with the description of two operations that can be performed on pairs of field elements. GF(2) is the smallest possible finite field, which contains only the integers 0 and 1 as field elements. Addition and multiplication are performed modulo 2, therefore addition is the logical XOR, and multiplication is the logical AND. A binary Galois field, $GF(2^m)$, comprises 2^m elements where m is a nonzero positive integer. Each element $a(t)$ in $GF(2^m)$can be uniquely represented with a polynomial of degree up to $m - 1$ with coefficients in $GF(2)$:

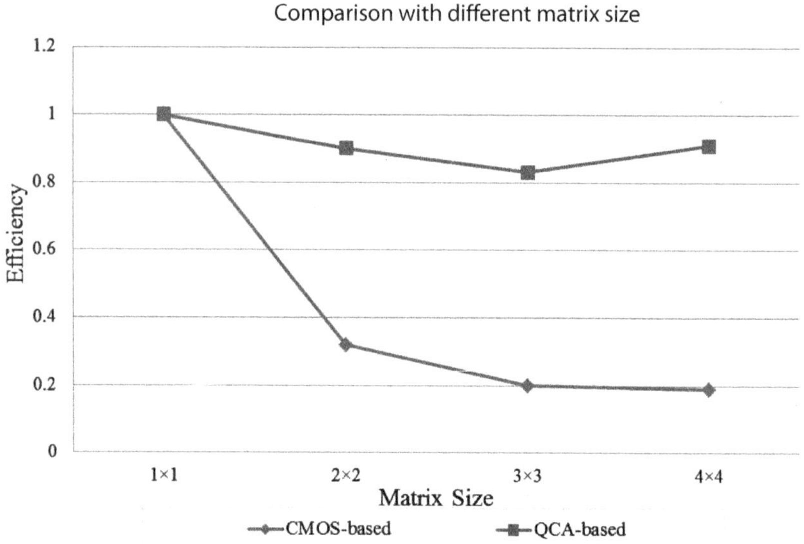

Figure 7.16 Efficiency versus matrix size.

$$a(t) = \sum_{k=0}^{m-1} a_t \cdot t^k = a_{m-1} \cdot t^{m-1} + \ldots + a_2 \cdot t^2 + a_1 \cdot t + a_0 \quad (7.3)$$

The following pseudo-code describes the algorithm for multiplying two polynomials $a(t)$ and $b(t)$, which belong to $GF(2^m)$ modulo an irreducible polynomial $p(t)$of degree m [12].

$r(t) = 0;$

for $i = m - 1$ *downto* 0 *do*

$\{r(t) = t \times r(t) + a_i \times b(t);$

if degree $(r(t)) = m$ then

$r(t) = r(t) - p(t);$

}

Return $r(t)$

7.4.2 QCA Systolic Galois Field Multiplier Design

In this section a QCA systolic array design of a $GF(2^m)$ multiplier is presented. The 4-bit $GF(2^m)$ systolic multiplier architecture is based on the design by Großschadl [12]. The DFG of this architecture is shown in Figure 7.17, where a_i, b_i, and p_i (the modulus) are the inputs. The circles marked R_i (the results) denote logic XOR operations, and the circles marked B_i and P_i indicate logic AND operations. According to this DFG, the serial inputs a_i and the feedback signals f_i have to reach the AND gates B_i and P_i at the same time, respectively. However, a direct mapping of this DFG to a QCA-based circuit is impossible due to the long wires. In addition, because of the internal delays of the QCA operations, it is impossible to get signals from B_i and P_i to R_i without any delay. To solve this timing issue, a retiming is performed, (i.e., delays are transferred or added on certain edges), to meet the timing requirements. In QCA circuit design, long wires are a problem, since they can affect the circuit stability. Therefore, the long wires are divided into different clocking zones under the timing requirements to ensure the robustness of the QCA circuit. In this design, QCA cut-set retiming, as proposed in Chapter 6, is applied.

The internal delay of each QCA component used in this design needs to be analyzed before applying the retiming technique. Two basic operations, namely two-input AND gates and three-input XOR additions, are used in this design. The AND gate is derived from a majority gate which requires at least one clocking zone delay (D^{-1}). A modified QCA CFA, as shown in Figure 7.18 is used to implement a three-input XOR operation. In this adder, two majority gates are cascaded. Hence, it requires at least two clocking zone delays (D^{-2}) from the inputs to the outputs. Therefore, the minimum delay for the three-input XOR operation is two clocking zone delays (D^{-2}).

In the QCA retiming procedure, first of all, a timing analysis is required to check whether the existing delays can satisfy the internal delay requirement

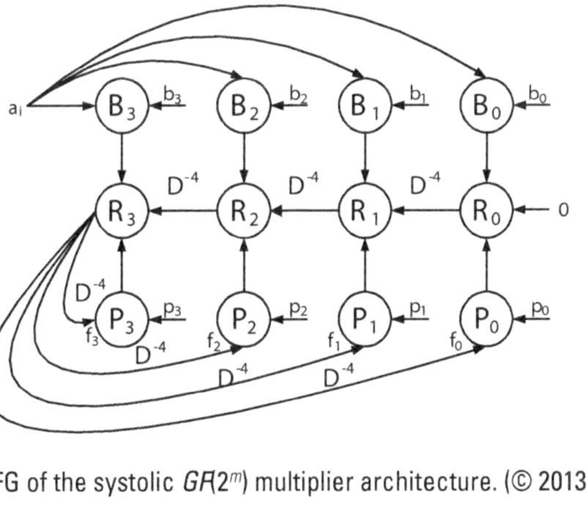

Figure 7.17 QCA DFG of the systolic $GF(2^m)$ multiplier architecture. (© 2013 IEEE. *From* [3].)

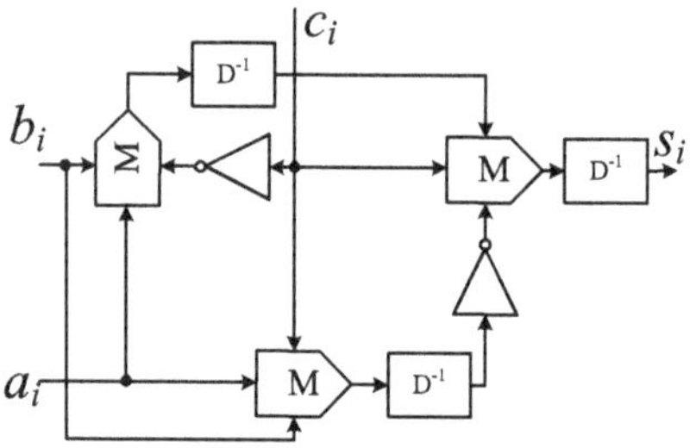

Figure 7.18 QCA three-input XOR schematic. (© 2013 IEEE. *From* [3].)

of each operation. To satisfy the minimum delay in the three-input XOR (R_i), two clocking zones delays (D^{-2}) need to be reserved along the outbound edges of R_i. As shown in Figure 7.17, D^{-4} delays have been allocated on R_i's outbound edges. This is sufficient to meet the timing requirements for this operation. Therefore, no cut is needed. As discussed earlier, the AND operation (B_i and P_i) requires a delay of at least D^{-1}. However, no delay exists on the current DFG. Therefore, Cut 1 is applied on the output of B_i as shown in Figure 7.19. An extra delay of D^{-1} is added on every outbound edge of this cut. Then the timing requirement of B_i is met. Similarly, the AND operation P_i has to meet the same timing requirement. Cut 2 is applied as shown in Figure 7.19. Note that this cut crosses both the inbound and outbound edges of the P_i operation. Hence, one delay, D^{-1}, is added on the outbound edge and a delay, D^{-1}, is reduced on the corresponding inbound edge to keep the timing relationship consistent.

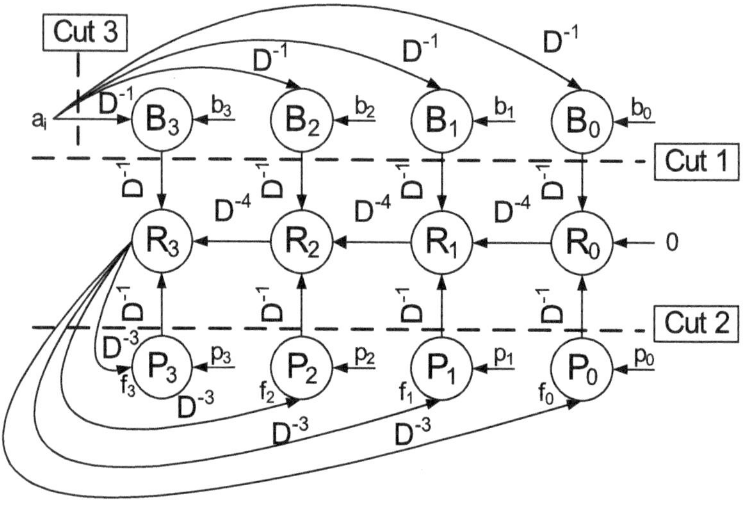

Figure 7.19 Retiming procedure: cut 1 to cut 3. (© 2013 IEEE. *From* [3].)

After sorting out the internal delays of each operation, the second step is to focus on the wire delays in the circuit. In the QCA circuit designs, long wires are partitioned into different clocking zones. At the top of the DFG, serial inputs a_i are required to reach B_i at the same time and without delay, which obviously violates the QCA timing constraint. Cut 3, as shown in Figure 7.19, is applied in this case to add an extra delay D^{-1} to these edges to ensure that at least a one clocking zone delay exists. However, the wires connecting a_i to B_0, B_1, B_2 are quite long, which may affect the robustness of this QCA circuit. In order to avoid any potential vulnerabilities, extra delays are required on these edges between a_i and B_i. Therefore, Cut 4, Cut 5, and Cut 6 are applied as shown in Figure 7.20. Delays are added on the outbound edge of B_i and subtracted from the corresponding outbound edges of R_i. From left to right, one clocking zone delay can be subtracted from the outbound edges of R_2, R_1, R_0 one by one. These delays are then added to the input edges. Similar delay transfers are applied at the bottom of the DFG to ensure sufficient delay on the feedback wires f_i. After these six cuts, it is clear that all QCA timing constraints are met and the original timing relationship of the design remains. The corresponding QCA circuit layout can be easily mapped from this re-timed DFG. Due to its fully pipelined nature, once data are input into the architecture, no extra control is required to manage the data path. The result is available directly after a certain number of clock cycles.

A schematic of the 4-bit QCA $GF(2^m)$ multiplier is shown in Figure 7.21 and its multilayer QCA design is shown in Figure 7.22. This design uses 20

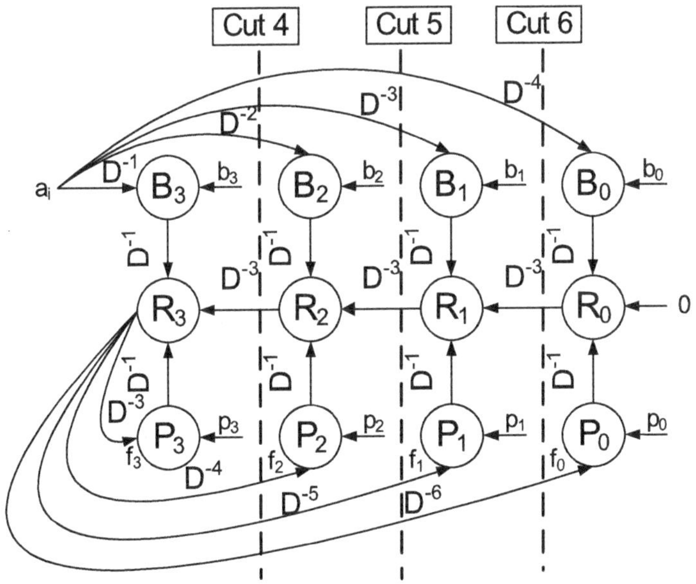

Figure 7.20 Retiming procedure: cut 4 to cut 6. (© 2013 IEEE. *From* [3].)

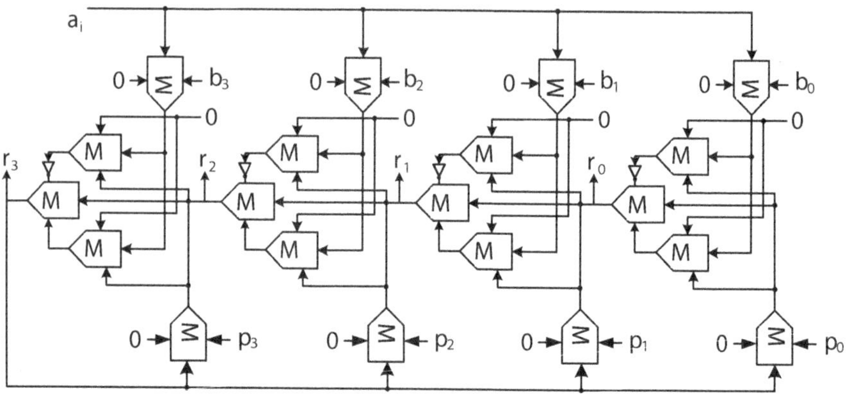

Figure 7.21 4-bit QCA $GF(2^m)$ multiplier systolic array schematic. (© 2013 IEEE. *From* [3].)

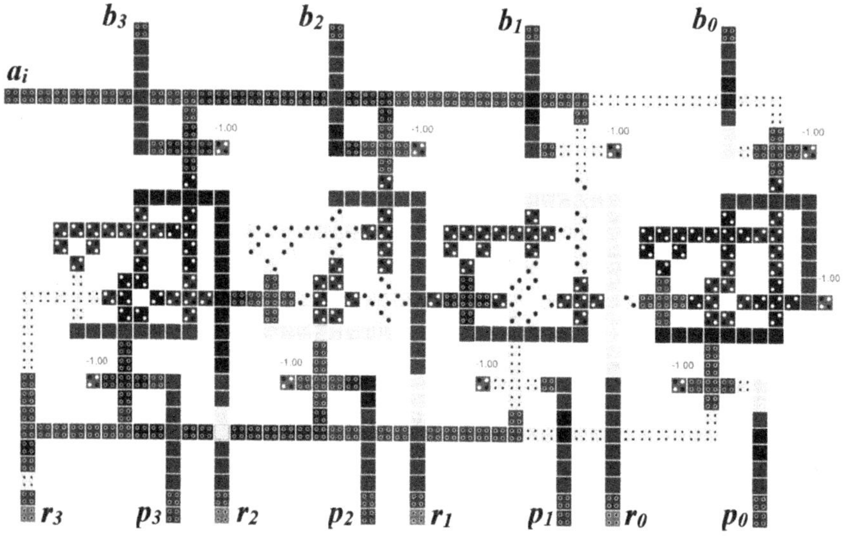

Figure 7.22 4-bit QCA $GF(2^m)$ multiplier multilayer systolic array layout. (© 2013 IEEE. *From* [3].)

majority gates and eight inverters. The data are fully pipelined and no control logic is required. Due to the pipeline nature of QCA technology, complex control logic, such as a FSM, would be more difficult to design since it involves more switches in the data paths. QCA systolic array designs have the advantage of minimum control logic. However, it is interesting to see the impact of the control logic on a QCA single processor (FSM-based) architecture with the same functionality. The following section provides designs and discussions of QCA single processor-based Galois field multiplier architectures.

7.4.3 QCA Single Processor Galois Field Multiplier Design

For the purpose of comparison, in this section, a QCA single processor multiplier for $GF(2^m)$ is designed using both multilayer and coplanar crossings. The basic block diagram of this design is divided into two parts, the computation module and the control module (FSM), as depicted in Figure 7.23. The computation module of this single processor architecture is the same as the PE in the corresponding systolic array architecture, and consists of two additions (B and P) and one three-input XOR operation (R). The internal delay of this computation module is four clocking zone delays, denoted as Z^{-1}. According to the $GF(2^m)$ algorithm, the output, f_i, is used as a feedback for the less significant bit calculation. In this single processor architecture, the output of the R operation could be either f_3, which is used as the input to P in the subsequent calculations, or f_2, f_1, f_0, which are used as one of the inputs of the next R calculation. Therefore, a more complicated control system is required for this single processor architecture to guarantee a correct data flow. The feedback signal f_i is split into two paths as shown in Figure 7.23. The bottom path is a feedback for the P operation. The most significant bit (MSB) computation result traverses this path. Since this data is used for all 4 bit calculations, the result, f_3, has to be held for at least four clock cycles. Due to the QCA wire's equivalence to a D-latch chain, a shift register architecture is the simplest design to provide this functionality. The bottom MUX is used to control whether the value of f_3 in the shift registers needs to be refreshed or remains. The first Z^{-1} delay is the internal delay of a 1-bit multiplexer. The top path is a feedback for the R operation. In the case of the 4-bit $GF(2^m)$ multiplier, this feedback is actually the value of f_2, f_1, f_0 or its initial value '0.' A simple shift register architecture is applied to ensure a correct

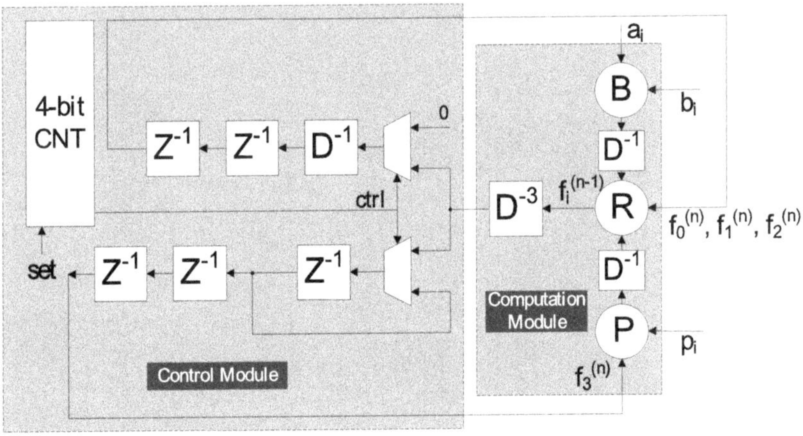

Figure 7.23 4-bit QCA single processor $GF(2^m)$ multiplier delay schematic. (© 2013 IEEE. *From* [3].)

data flow. The MUX in this top data path has only a single clocking zone delay (D^{-1}), due to its much simpler architecture, in which one input is a '0.' Both multiplexers are controlled by a 4-bit counter. This counter is implemented as a loop of QCA wire. Once *set* becomes valid, a datum circulates in the loop and outputs every 4 clock cycles.

The QCA gate-based schematic and circuit layouts of this single processor architecture for a 4-bit $GF(2^m)$ multiplier are shown in Figures 7.24–7.26. In Figure 7.25, on the left side of the control module is the 4-bit counter. The counter consists of a majority gate with a four clock cycle delay wire loop. Once the counter is triggered by setting the *set* signal to '1,' this '1' circulates in the loop and is output once every four clock cycles for use as the selection signal for the multiplexers. Two ovals highlight the feedback wires (i.e., w_1 and w_2), which also function as shift registers. The control module requires a lot of space in this layout. As the data word size of the multiplier increases, it is anticipated that the size of the control module will increase while the computation module will remain unchanged. Based on the schematic, three parts of the control module change when the word size doubles. First, to maintain correct data flow, [i.e., the input of correct feedback information (f_i) to the computation module], the length of the bottom feedback wire (w_2) doubles. Similarly, the top feedback wire (w_1) length will be one Z^{-1} less than w_2. Secondly, the size of the counter needs to double when the word size of the multiplier doubles. However, because this is a single processor architecture, the computation module only calculates 1 bit of information. Therefore, its size remains the same. These trends will be discussed in the following subsection. The coplanar design performs similarly to the multilayer design. In this single processor design, only two crossings are

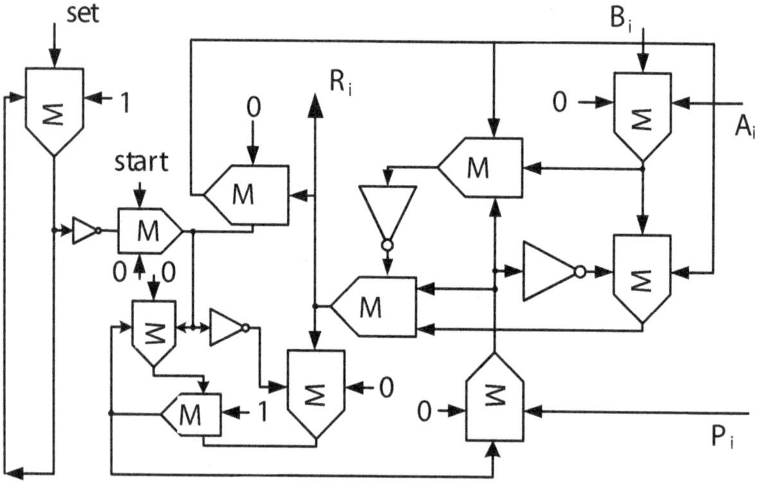

Figure 7.24 Single processor 4-bit QCA $GF(2^m)$ multiplier schematic. (© 2013 IEEE. *From* [3].)

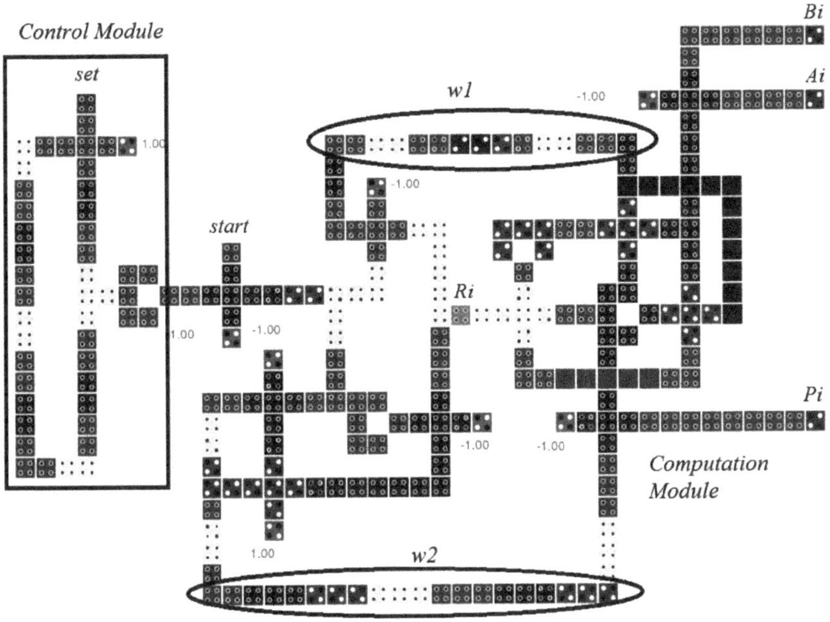

Figure 7.25 Single processor 4-bit QCA $GF(2^m)$ multiplier multilayer layout. (© 2013 IEEE. *From* [3].)

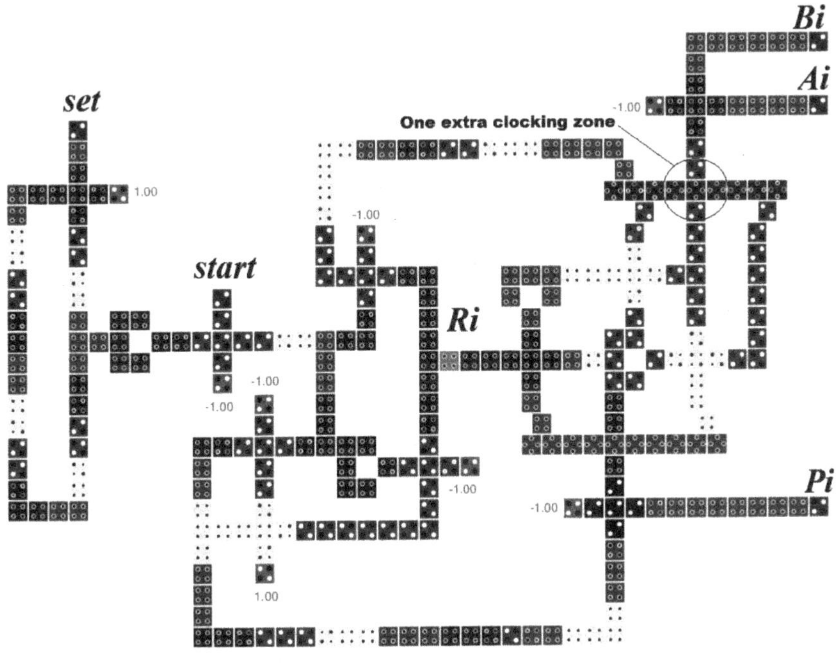

Figure 7.26 Single processor 4-bit QCA $GF(2^m)$ multiplier coplanar layout. (© 2013 IEEE. *From* [3].)

required. Therefore, after carefully placing the clocking zones, only one extra clocking zone delay is introduced in the coplanar design, as shown in Figure 7.26, when compared to the multilayer design.

7.4.4 Design Study

7.4.4.1 Simulation and Results

The $GF(2^m)$ multipliers in both the systolic array and single processor architectures in the previous section were simulated using QCADesigner. The same simulation parameters were used as in case study I.

For a 4-bit $GF(2^m)$ multiplication, the multilayer and coplanar systolic array designs take 5 and 5.5 clock cycles to achieve a correct result, respectively. Meanwhile, the multilayer and coplanar single processor architectures take 17 and 17.25 clock cycles to achieve the correct result. Therefore, once again the advantages of using QCA systolic array design in terms of its data parallelism and pipelining to increase the circuit speed is evident.

7.4.4.2 Analysis and Comparison

In this case study, the result from a previous PE is used as the input for the next PE in the systolic array. As discussed earlier, the processor requires global feedback. Therefore, the data flow management becomes particularly important in designing the processor. The difficulty and cost in designing the circuit is discussed in this section. For the purpose of comparison, a single processor design in both CMOS and QCA technology has been implemented. The resources used in the QCA designs are listed in Table 7.6. The inverter and wire cell counts in this table are provided for both the MC and CC designs. All design results are shown in Table 7.7.

QCA is a technology based on nano scale cells, which results in a very small fabrication size. To offer a fair comparison with CMOS, the control area increase ratio is defined. This indicates the increase in control area over the smallest case, namely the 4-bit $GF(2^m)$ multiplier implementation in the corresponding technology. The multilayer and coplanar QCA designs have the same control module areas because there are no wire crossings in the control circuits.

Table 7.6
Single Processor Galois Field Multiplier Resource Count

Designs	Number of M	Number of Inv MC/CC	Number of Crossings	Number of Wire Cells MC/CC
4-bit	11	4/9	2	203/193
8-bit	11	4/9	2	285/273
16-bit	11	4/9	2	477/465
32-bit	11	4/9	2	861/849

Table 7.7
Single Processor Design Cost Comparison

	CMOS (65 nm)			**QCA-Multilayer (20 nm)**			**QCA Coplanar (20 nm)**		
	Total (μm^2)	**ctrl (μm^2)**	**Ratio**	**Total (μm^2)**	**ctrl (μm^2)**	**Ratio**	**Total (μm^2)**	**ctrl (μm^2)**	**Ratio**
4-bit	83.8	76.1	N/A	0.44	0.30	N/A	0.46	0.30	N/A
8-bit	130.9	123.2	0.62	0.68	0.54	0.80	0.68	0.54	0.80
16-bit	210.6	202.9	1.67	1.21	1.07	2.57	1.21	1.07	2.57
32-bit	368.0	360.3	3.73	2.51	2.37	6.90	2.51	2.37	6.90

The ratio trends for this case are shown in Figure 7.27. The area cost of the control circuit is significant in the QCA designs. This is because the feedback wire lengths increase significantly for the larger word size implementations. Zig-zag-shaped QCA wires are used to save space, but a safe space between wires has to be included. Therefore, after optimization, the area cost of the control circuit of the QCA architecture still dominates the total cost, and this cost increases significantly as the word size increases. The impact of the control system grows much faster than that in CMOS technology. A finite state machine (FSM) is difficult to design and very costly in QCA technology. Systolic arrays can be used to reduce the size of the control logic, which greatly benefits QCA circuit design and minimizes the impact of the extra area caused by QCA feedback wires in the control circuit. This leads to the overall conclusion that systolic array architectures are particularly attractive in QCA circuit design.

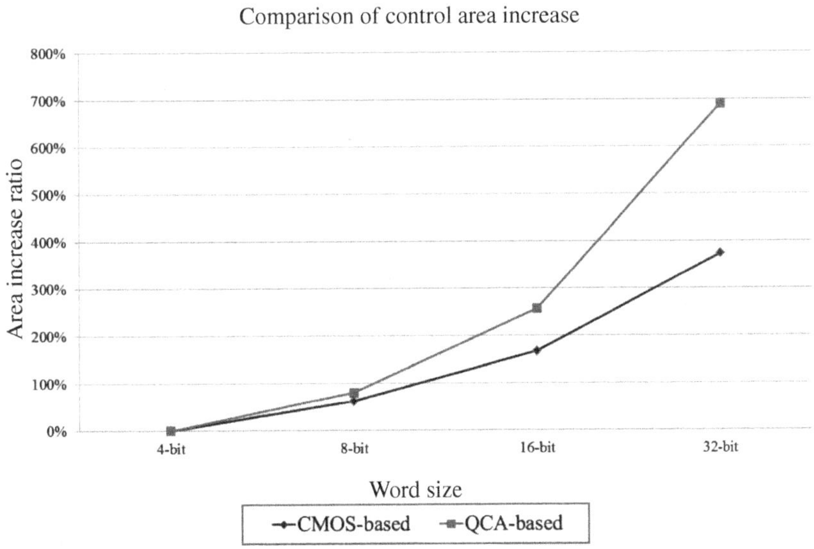

Figure 7.27 Control circuit cost comparison. (© 2013 IEEE. *From* [3].)

7.5 QCA Systolic Array Design Methodology

Based on the two case studies discussed above, in this section, a general QCA systolic array design methodology has been developed. This methodology consists of five steps, described as follows.

- Step 1: *Algorithm analysis.* In this step, similar to CMOS architecture design, the target algorithm is analyzed to ensure that it has a data flow that can be pipelined. Then a PE is designed that can be replicated in a systolic array. If the outputs of each PE are independent, as in case I, the main design load will be focused on the QCA logic design. If the results of the PEs are codependent, then it is essential to analyze the timing relationships after the basic logic design is complete to ensure that they meet the QCA timing requirements.
- Step 2: *Logic design.* In this step, the algorithm is first transformed to majority gate and inverter based logic. Optimization methods are applied to reduce the number of gates and to minimize the circuit size, [13]. As discussed earlier, if the logic does not satisfy the requirements, further timing analysis is required. Note that a small zone region floorplan as described in Chapter 2 is used in this research. If a columnar floorplan is desired, the clocking zone floorplan has to be designed in conjunction with the logic design process, which might result in a different circuit layout.
- Step 3: *Timing analysis.* This step is critical in QCA circuit design. Existing delays in the original architecture may not be sufficient if mapped directly to QCA. Therefore, delays need to be adjusted to satisfy the QCA timing rules. QCA cut-set retiming, as proposed in Chapter 6, is a very useful technique to resolve these timing issues. Scaling, inserting, and transferring of delays should be carried out as appropriate. Timing issues may require multiple passes of cut-set retiming, and the process may need to be applied again once the layout design is complete to ensure correct timing. When designing a QCA circuit using coplanar crossings, the delays at the crossings need to be analyzed in this step. If both of the wires at a crossing are in the same clocking zone, then an extra clocking zone delay needs to be added to the regular cell wire to ensure proper signal propagation. Otherwise, no extra delays are required for this coplanar crossing.
- Step 4: *Layout.* In this step, the QCA layout is mapped from the retimed architecture. It has to be noted that sometimes new overlong wire delays will be introduced during the layout design, which will create timing

issues that were not recognized at the architecture level. Therefore, the timing analysis step may need to be repeated.

- Step 5: *Verification.* When all the timing constraints are met, the functionality of this systolic array design can be verified by QCA tools, such as QCADesigner.

7.6 Conclusion

Two QCA systolic array architectures are presented in this chapter as case studies to demonstrate important QCA design characteristics. It is shown that the characteristics of systolic arrays, such as pipelines, parallelism, and simple control, are well-suited to QCA technology. These advantages are further magnified when compared with corresponding CMOS implementations.

From the case studies, it can be concluded that when a pipelined single processor architecture (all results are independent from each other) is converted to a systolic array architecture, the disadvantage of the internal delay of the QCA circuit is reduced in the systolic array design due to its inherent parallelism. The data path of a single processor architecture with global feedback cannot be pipelined, and a large control unit is required. This type of architecture goes against QCA's pipeline nature. Therefore, to design such a circuit, QCA technology requires significant resources to build a complex control system. This disadvantage is magnified when compared with CMOS technology. However, conversion to a systolic architecture, not only improves the computation speed, but also greatly simplifies the control logic. A fully pipelined systolic array does not need extra control logic to manage the data flow. Although, the quantitative results of the research outlined in this chapter may vary for other forms of QCA, the principle advantages of the proposed QCA systolic array methodology would still apply.

References

[1] Lu, L., et al., "QCA Systolic Matrix Multiplier," in *Proceedings of the IEEE Computer Society Annual Symposium on VLSI,* 2010, pp. 149–154.

[2] Liu, W., et al., "Design Rules for Quantum-dot Cellular Automata," in *Proceedings of the IEEE International Symposium onCircuits and Systems,* 2011, pp. 2361–2364.

[3] Lu, L., et al., "QCA Systolic Array Design," *IEEE Transactions on Computers,* Vol. 62, 2013, pp. 548–560.

[4] Zhang, R., et al., "Performance Comparison of Quantum-dot Cellular Automata Adders," in *Proceedings of the IEEE International Symposium on the Circuits and Systems,* 2005, pp. 2522–2526.

[5] Cho, H., and E. Swartzlander, Jr., "Adder Designs and Analyses for Quantum-Dot Cellular Automata," *IEEE Transactions on Nanotechnology,* Vol. 6, 2007, pp. 374–383.

[6] Cho, H., and E. Swartzlander, Jr., "Adder and Multiplier Design in Quantum-Dot Cellular Automata," *IEEE Transactions on Computers,* Vol. 58, 2009, pp. 721–727.

[7] Parhi, K., *VLSI Digital Signal Processing Systems: Design and Implementation,* New York: Wiley, 1999.

[8] Kung, H., and C. Leiserson, "Systolic Arrays (for VLSI)," in *Sparse Matrix: Proceedings 1978,* 1978, pp. 256–309.

[9] Walus, K., et al., "QCADesigner: A Rapid Design and Simulation Tool for Quantum-Dot Cellular Automata," *IEEE Transactions on Nanotechnology,* Vol. 3, 2004, pp. 26–31.

[10] Gin, A., P. Tougaw, and S. Williams, "An Alternative Geometry for Quantum-Dot Cellular Automata," *Journal of Applied Physics,* Vol. 85, 1999, pp. 8281–8286.

[11] Lent, C., and P. Tougaw, "A Device Architecture for Computing with Quantum Dots," *Proceedings of the IEEE,* Vol. 85, 1997, pp. 541–557.

[12] Großschadl, J., "A Low-Power Bit-Serial Multiplier for Finite Fields *GF* (2^m)," in *Proceedings of the IEEE International Symposium on Circuits and Systems,* Vol. 4, 2001, pp. 37–40.

[13] Zhang, R., et al., "A Method of Majority Logic Reduction for Quantum Cellular Automata," *IEEE Transactions on Nanotechnology,* Vol. 3, 2004, pp. 443–450.

8

Evaluation of QCA Circuits with New Cost Functions

Weiqiang Liu, Liang Lu, Máire O'Neill, and Earl E. Swartzlander, Jr.

8.1 Introduction

To date, circuit designs in QCA have been extensively studied due to the development of suitable simulation tools [1]. Arithmetic circuits [2–4], memory [5, 6], and simple processors [7, 8] have been designed and analysed. Digital design methodologies for QCA circuits have also been explored to achieve more efficient designs [9]. However, how to properly evaluate QCA circuits and compare them fairly has not been carefully considered in the research to date.

Among the circuits designed in QCA, adders have received considerable interest [10–15] and are the most extensively studied QCA circuits due to their importance in a computing system. Moreover, binary arithmetic designs played an important role in the development of CMOS cost functions [16–18]. Therefore, QCA adder designs have been selected as representative QCA circuits to be evaluated with the new cost functions proposed in this chapter.

As addition is at the heart of computer arithmetic, designers would like to use the best adder in their designs, but what defines the "best" adder in QCA? Previous research has found that multilayer QCA adders generally outperform coplanar adders according to metrics such as area, latency, and the number of cells [12, 15], which are directly mapped from CMOS metrics. However, the area in nanoscale integration is not as important as it used to be. Besides, the area advantage of multilayer adders is attained by using multilayer crossovers that are costly to fabricate and the difference between the two types of crossings in QCA is not included in the comparison metrics. Therefore, the previous metrics are not sufficient in providing a fair comparison due to the fundamental differences between QCA and CMOS technologies and QCA's unique characteristics. New metrics and cost functions for QCA circuits need to be investigated to help guide the optimization of QCA circuit design, which will contribute to the understanding of QCA technology.

In this chapter, the evolution of the cost metrics used in CMOS technology is revisited to provide clues on appropriate QCA metrics. The metrics used in previous research are reviewed and several new metrics are discussed. Based on the analysis of all available metrics, a family of new cost functions is proposed. The proposed cost functions can be used to evaluate any QCA design. Adder designs are selected as representative designs for evaluation in this work due to the reasons mentioned above. A representative number of QCA adder designs are reviewed, and general metric formulas for each adder derived based on its schematic and layout. The QCA adders are compared in terms of the important metrics and then evaluated with the proposed cost functions in terms of the overall cost. The comparison results show that the selection of the "best" adder is dependent on both the design goals and the implementation technology. This work is based on semiconductor QCA, in which each cell can be blocked individually. Further work on metrics and cost functions would report a more detailed analysis of clock structure to show its effects on performance. This work is the first to investigate the overall cost of QCA circuit design, and it is anticipated that it will inspire further consideration of cost functions for future QCA designs. The continued development of QCA nanotechnologies will inevitably lead to more accurate cost functions in the future.

The remainder of this chapter is organized as follows. Section 8.2 proposes several comparison metrics and studies the new cost functions. Section 8.3 provides an overview of six representative adder designs in QCA. General formulas for the metrics are derived. Section 8.4 summarizes the formulas for QCA adders and compares them according to each metric. Then the comparison results based on the proposed cost functions are provided and discussed. Conclusions are given in Section 8.5.

8.2 QCA Cost Metrics and Cost Functions

The design trade-offs between area, delay, and power consumption are usually used in cost functions to measure the quality of a CMOS circuit.

In the past, the area of a CMOS circuit was of primary concern as the die size had a strong effect on the yield and, thus, the cost. As the feature size has been reduced significantly due to advances in CMOS technology, this metric is no longer as significant as it used to be. With the advance of technology, the main goal of digital design changed to high speed and performance. Therefore, area-delay products became the dominant cost functions used in CMOS. Cost functions based on the area-time product were shown to be useful to find an optimum memory design in very-large-scale integration (VLSI) by Mead and Rem [19]. Models of the area-time trade-off were further studied by Thompson [20, 16], who proposed the following:

$$\text{Cost}_{\text{Area-Delay}} = A \times T^n, 0 \le n \le 2 \tag{8.1}$$

where A is the area and T is the delay of a CMOS circuit. Lower bounds of these cost functions were derived from the theory of the bisection problem based on Mead and Rem's VLSI model [19]. Brent and Kung also showed the bounds of similar cost functions for binary arithmetic [17]. In both models, common assumptions are made, including that the wires and nodes should have a minimum width and spacing, λ, determined by the manufacturing minimum feature size and that it takes a unit of time to transmit information over a wire. The cost functions proposed are valid for all VLSI circuits that satisfy the assumptions. The models have been well accepted in CMOS technology. Similarly, for QCA there is a minimum manufacturing width for a QCA wire, which is the size of a QCA cell (similar to the λ-rule in CMOS) [21]. There is also a minimum unit of time needed to transmit information in QCA wires, which is a clocking zone delay (equal to a quarter of a clock cycle). This suggests that the models can also be applied to QCA technology. Although the exact lower and upper bound of the area-time product for QCA circuits may be different, there are also area-time trade-offs in QCA circuits.

Currently, power consumption is also very important, which is attributed to both increasing levels of integration and the demands of mobile computing. Transistor sizing can improve the speed of a circuit while increasing the power dissipation at the same time. Therefore, cost functions that involve power-delay products have become popular [18, 22] as follows:

$$\text{Cost}_{\text{Power-Delay}} = P^m \times T^n \tag{8.2}$$

where P is the power dissipation and largely determined by dynamic power dissipation in CMOS. The dynamic power dissipation is given by:

$$P_{\text{Dynamic}} = \alpha C_L V_{dd}^2 f \tag{8.3}$$

where α is the activity factor, C_L is the load capacitance, V_{dd} is the supply voltage and f is the clock frequency. Once the supply voltage and clock rate are determined, the power dissipation of a CMOS circuit is mainly dependent on α and C_L. Different algorithms and architectures significantly affect the activity factor (α) of a circuit [23]. The power dissipation should also be considered in QCA cost function designs. As irreversible power dissipation (which is negligible in CMOS) dominates the total power dissipation in QCA, the power dissipation issue is different from that of CMOS, which will be further discussed in this section.

Similar to CMOS, the cost metrics for QCA circuits need to be carefully investigated as these can significantly affect the choice of QCA circuit design. Previously, researchers have used the number of cells (also referred to as complexity), area, and delay as the metrics to compare different QCA circuits. In this section, these metrics will be reviewed. Some new metrics including the number of logic gates, the irreversible power dissipation, and the number of crossovers are investigated. Furthermore, a family of new cost functions is proposed.

8.2.1 Area/Complexity

The number of cells (sometimes referred to as complexity) in QCA circuits is analogous to the number of transistors in CMOS circuits. In a QCA circuit, both logic components and connecting wires are composed of cells. As a result, the number of cells in a QCA circuit is generally proportional to its area [15, 24], and including both would result in a double weighted area metric. Therefore, it seems inappropriate to include both in a QCA cost function. Either the number of cells or the area can be used as a rough measure of the complexity of a QCA circuit. As the feature size of a QCA cell is even smaller than that of CMOS circuits (as small as 1 nm for atomic and molecular QCAs), the number of cells and the area of a QCA circuit are not as important as other metrics. Therefore, these two metrics can be included in a QCA cost function with very low priority. To measure the complexity of a QCA circuit, the numbers of logic gates and crossovers are better metrics. The circuit complexity in QCA is actually the sum of the three primitives: majority gates, inverters, and crossovers. In this work, the number of logic gates and crossovers are used to measure the complexity of a QCA circuit.

8.2.2 Delay

Delay is always an important metric in assessing the performance of circuits. It is assumed that the various circuits in this work will be clocked at the same rate, so the number of clocking zones serves as a measure of the latency. The delay of any QCA circuit is determined by the clock period times the number of clocking zones along its critical path. Better QCA circuit designs use fewer clocking zones. Therefore, the delay of a QCA circuit should be included in a cost function. In CMOS technology, delays are compared in terms of clock cycles, while in QCA, since the minimum delay in QCA is a clocking zone delay, QCA delays are compared in units of a quarter of a clock cycle.

8.2.3 Irreversible Power Dissipation

Power dissipation has been a limiting factor in modern integrated circuits for a long time. Although the power dissipation in QCA is small, a QCA circuit could still incur thermal problems due to the extremely high density. According to Landauer's principle [25], which has recently been experimentally verified [26], when the internal entropy of a physical system is decreased during computations, heat must be expelled from the system. Although this "irreversible dissipation" is negligible in traditional technology, it becomes a major limitation in computers with ultra-high density nanoscale integration [27, 28].

QCA "wires," which are composed of cells, can be considered as shift registers. As there is no information loss in this shift register, it dissipates little power [27]. Similarly, the inverter gate is also a logically reversible unit, that consumes little power. The irreversible dissipation mostly comes from the three-input majority gate, since the three-to-one unbalanced mapping produces significant information loss and unavoidable power dissipation. For equiprobable inputs, the more majority gates used in a QCA circuit, the more power that is likely to be consumed, which is equivalent to the activity factor (α) in (8.3). Different algorithms and architectures use different numbers of majority gates, which leads to different activity factors as discussed above. Therefore, majority logic reduction methods should be employed in QCA designs [15, 29, 30].

8.2.4 Number of Crossovers

A critical issue in QCA technology is the crossover of two separate signal wires. The unique coplanar crossings use only one layer and require precise alignment during fabrication [31]. An alternative geometry is the multilayer crossover. The original idea of multilayer QCA circuits was proposed by Gin et al. [32] and the idea was further developed in QCADesigner [33] which applies multilayer crossovers to large QCA designs. In [32], a biplanar fabrication scheme using two small silicon layers inside a SiGe device was proposed to implement noncoplanar QCA cells. Implementing this kind of noncoplanar cells is more difficult than implementing coplanar cells. Furthermore, this biplanar proposal can only be used for two-layer QCA circuits. For a multilayer crossover as proposed in QCADesigner, at least three layers are required to implement the crossings, and the distance between two vertical neighboring layers needs to be arranged properly to match the kink energy to that of normally adjacent cells, which may be difficult to achieve. As a result, significantly more complexity is required to fabricate multilayer crossovers compared with coplanar crossings. While a number of single-layer experiments have been demonstrated [34, 35], there have been no prototypes of multilayer QCA designs manufactured to date.

The clocking of both crossover types is a challenge during operation. For coplanar crossings, very fine clocking zones are required for robust signal propagation. The clocking of multilayer crossovers is also expected to be difficult. Thus, when more crossovers are used in a QCA circuit, it will be more difficult to fabricate the circuit [36].

As discussed above, the cost of a multilayer crossover is greater than that of a coplanar crossing, which suggests a cost model as follows:

$$C_{ml} = m \times C_{cp} \tag{8.4}$$

where, C_{ml} is the cost of a multilayer crossover, C_{cp} is the cost of a coplanar crossing, and m is a coefficient to reflect the higher cost of a multilayer crossover over a coplanar crossing. As a multilayer crossover uses at least three single layers, m is assumed to be three or more in this work.

The crossovers in QCA are important physical implementation constraints in realizing complex circuits. Minimizing the number of crossovers is always desirable in QCA circuit design [36]. Thus, the number of crossovers is a unique and very important metric in QCA that should be included in QCA cost functions.

8.2.5 Proposed QCA Cost Functions

Based on the above discussion, it is believed that either the number of cells or the area can be included in a cost function with very low priority. To measure the complexity of a QCA circuit, the numbers of logic gates and crossovers are better metrics. Furthermore, the majority gates are associated with irreversible power dissipation, and crossovers are associated with fabrication difficulty. Therefore, the number of majority gates and crossovers should have higher weightings. Delay is always important due to the performance considerations. For these reasons, the delay, number of logic gates and number of crossovers are used to measure the performance, complexity, irreversible power dissipation and the fabrication difficulty of a QCA circuit. A generalized QCA cost function is proposed as follows:

$$\text{Cost}_{QCA} = \left(M^k + I + C^l\right) \times T^p, 1 \le k, l, p \tag{8.5}$$

where M is the number of majority gates, I is the number of inverters, C is the number of crossovers ($C_{ml} = m \times C_{cp}$), T is the delay of the circuit, and k, l, p are the exponential weightings for majority gate count, crossover count, and delay, respectively. A constant weighting of '1' is assigned to the number of inverters, as inverters only affect the complexity of QCA circuits. These cost functions are similar to the CMOS area-delay cost functions as expressed in (8.1). However, the major differences in complexity, the irreversible power, and crossover types between QCA and CMOS are distinguished here. Therefore, M_k and C_l are included in the cost functions to reflect their effect on QCA design.

The cost functions prioritize different metrics according to the weightings k, l, and p. For example, if speed is a primary concern, more weight can be given to the delay metric, (i.e., a higher value of p). If fabrication cost is more important, the value of l should be higher than that of p and k and so on. Therefore, the weight values can be adjusted depending on the overall design optimization goal.

8.3 Overview of QCA Adders[1]

Adders are at the heart of computer arithmetic, which has been extensively studied in QCA. In this section, six representative adders which include both coplanar designs and multilayer designs are compared and analyzed with the proposed cost functions. Coplanar adders designed to date are all RCA and include the first QCA adder [10] and optimized coplanar adders [11, 13]. Multilayer adders include the first majority logic reduced RCA [12], a CFA [14] and the Brent-Kung adder (a parallel prefix adder) [15] with the latter claiming the best performance to date.

8.3.1 Coplanar Adders

8.3.1.1 Tougaw Adder

The first QCA FA was proposed by Tougaw [10]. It uses five majority gates and three inverters. The FA is expressed as follows:

$$s_i = M\left[M\left(\bar{a}_i, b_i, c_{i-1}\right), M\left(a_i, \bar{b}_i, c_{i-1}\right), M\left(a_i, b_i \bar{c}_{i-1}\right)\right] \tag{8.6}$$

$$c_i = M\left(a_i, b_i, c_{i-1}\right) \tag{8.7}$$

These 1-bit FAs can be easily chained to produce an n-bit ripple carry adder, which is referred to as Tougaw adder in this chapter (the same naming convention is used in the rest of this chapter). A schematic of a 4-bit Tougaw adder is shown in Figure 8.1. It can be easily inferred that the number of majority gates, inverters and crossovers are $5n$, $3n$, and $9n$, respectively for an n-bit Tougaw adder. As the original design is not clocked, the same clocking zone assignment as used in Wang's adder [11] can be applied to the Tougaw adder. Figure 8.2 shows a layout of the 4-bit Tougaw, which is chained diagonally with

1. Section 8.3 is based on [37].

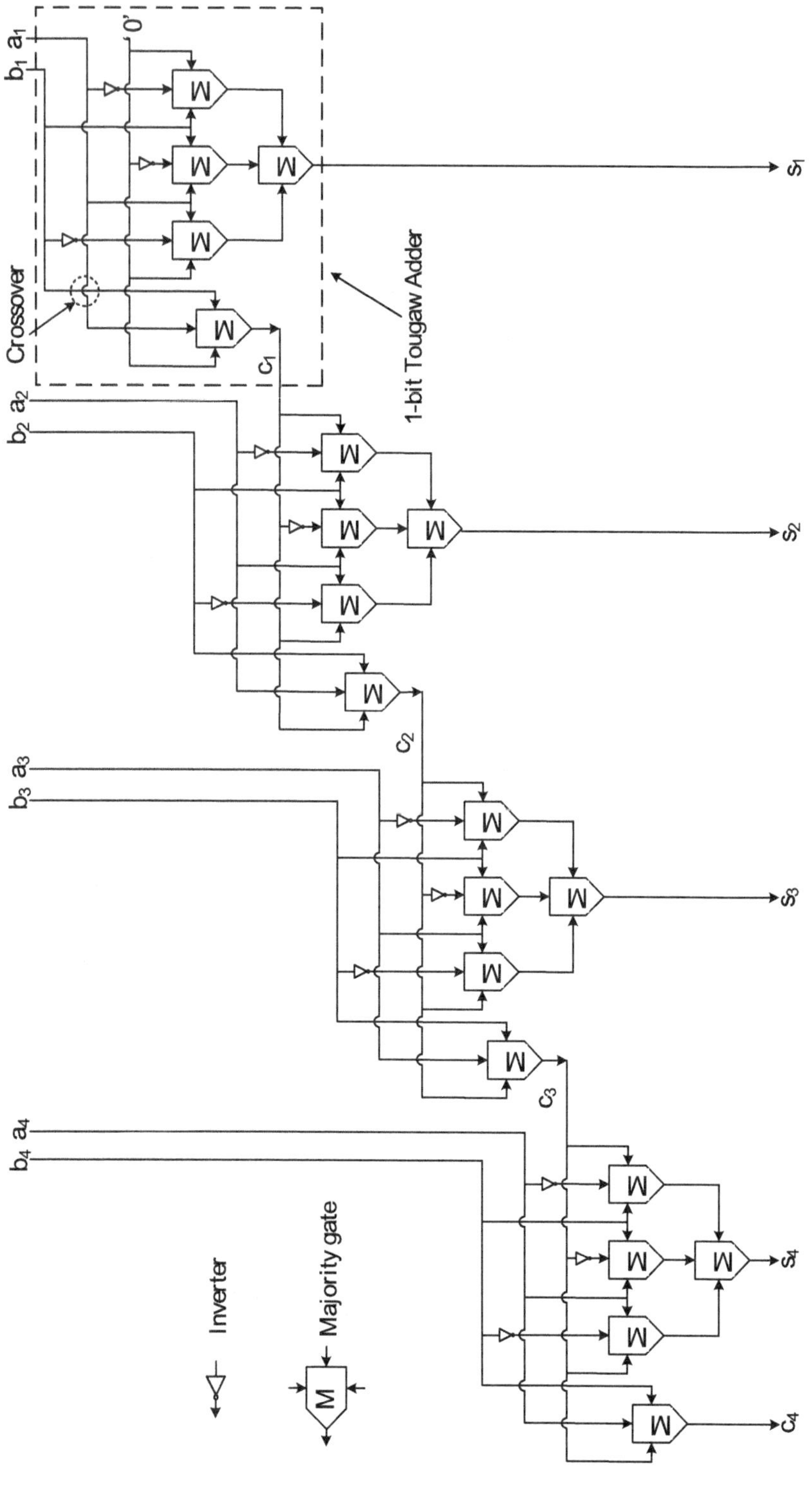

Figure 8.1 Schematic of a 4-bit Tougaw adder. (© 2012 IEEE. *From* [37].)

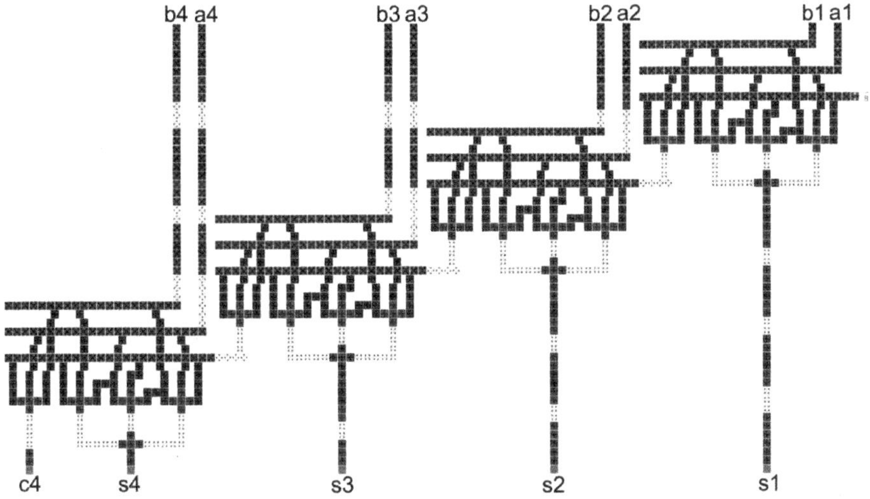

Figure 8.2 Layout of a 4-bit Tougaw adder. (© 2012 IEEE. *From* [37].)

1-bit Tougaw adders to accommodate the wire delays. From the layout, it can be easily derived that the delay for an n-bit Tougaw adder is $n+\frac{1}{4}$ clock cycles. Note that all the layouts in this chapter are shown to the same scale, so the size of the figures indicates the relative size of the adders.

8.3.1.2 Wang Adder

An optimized design of a 1-bit FA was proposed by Wang [11], in which majority logic reduction is used to simplify the design. Wang's adder uses only 3 majority gates, 2 inverters, and 6 crossovers. Therefore the complexity of the adder is significantly reduced. The calculation of the carry-out is the same as that of the Tougaw adder. However, the sum bit is optimized as follows:

$$s_i = M\left[\overline{c}_i, c_{i-1}, M\left(a_i, b_i, \overline{c}_{i-1}\right)\right] \tag{8.8}$$

The schematic and layout of a 4-bit Wang adder are shown in Figure 8.3 and Figure 8.4, respectively. An n-bit Wang adder consists of $3n$ majority gates, $2n$ inverters and $6n$ crossovers. Although the complexity is reduced, the delay is the same as that of the Tougaw adder, which is $n+\frac{1}{4}$ clock cycles for an n-bit Wang adder.

8.3.1.3 Hänninen Adder

The Hänninen adder [13] uses the same logical structure as Wang's adder but with an optimized layout. The clocking zones are rearranged to achieve a ro-

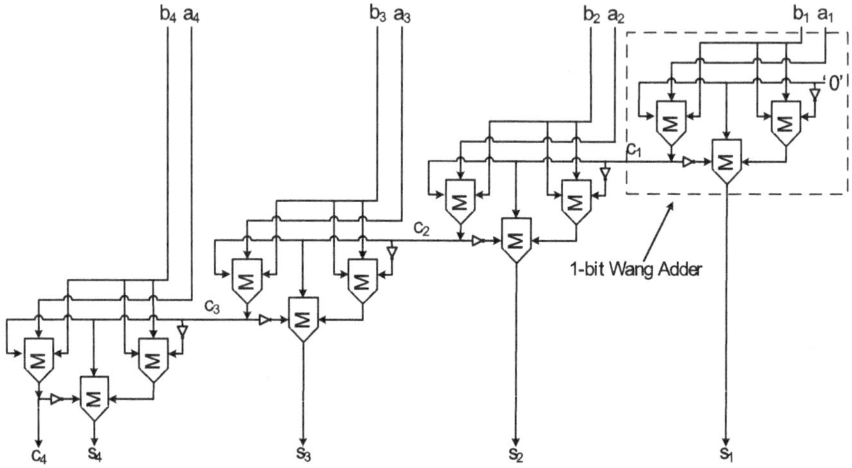

Figure 8.3 Schematic of a 4-bit Wang adder. (© 2012 IEEE. *From* [37].)

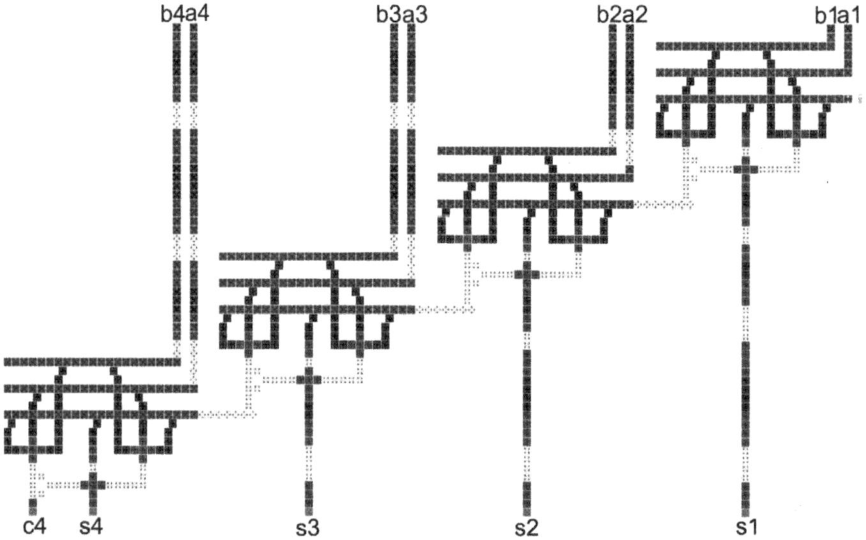

Figure 8.4 Layout of a 4-bit Wang adder. (© 2012 IEEE. *From* [37].)

bust design. The schematic and layout of a 4-bit Hänninen adder are shown in Figure 8.5 and Figure 8.6, respectively. The number of majority gates, inverters and crossovers of an n-bit Hänninen adder are $3n$, $2n$ and $3n$, respectively, which can be easily derived from its schematic. The delay for an n-bit adder is $n + 1$ clock cycles.

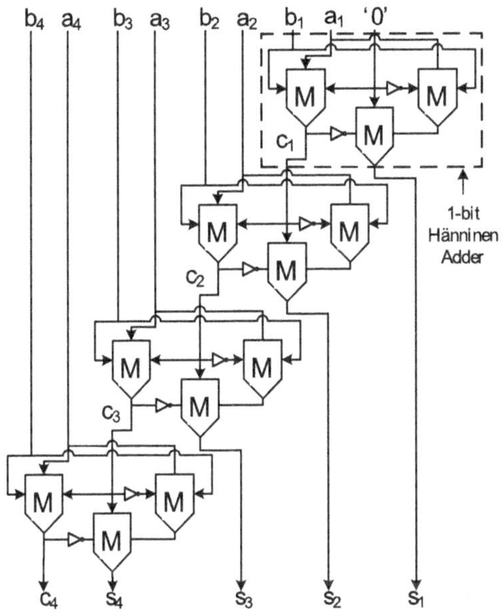

Figure 8.5 Schematic of a 4-bit Hänninen adder. (© 2012 IEEE. *From* [37].)

8.3.2 Multilayer Adders

8.3.2.1 Zhang Adder

Wang's adder was revised for multilayer implementation by Zhang [12]. The revised design uses the same addition structure. However, different types of crossovers are used. Its schematic is shown in Figure 8.7. The layout is designed using multilayer crossovers, which is shown in Figure 8.8. As the structure is the same as Wang's, the number of majority gates and inverters for an n-bit Zhang adder is also $3n$ and $2n$, respectively. However, the number of crossovers and the delay of an n-bit Zhang adder are different, $3n$ and n clock cycles, respectively. These can be derived easily from its layout. Although the number of crossovers used here is less than that in Wang adders, the cost of coplanar and multilayer crossings is quite different. The effect of the crossover cost will be further discussed in Section 8.4.

8.3.2.2 Cho Adder

The preceding RCAs all have four clocking zone delays per bit. A CFA was proposed by Cho [14], which is a layout optimized multilayer FA. In QCA the path from carry in to carry out uses one majority gate, which requires one clocking zone per bit in a RCA. The QCA CFA (referred to as the Cho Adder in this chapter) consumes only one clocking zone delay per bit, which significantly reduces the delay for large adders. The schematic and layout of a 4-bit Cho adder are shown in Figures 8.9 and 8.10, respectively. For an n-bit Cho

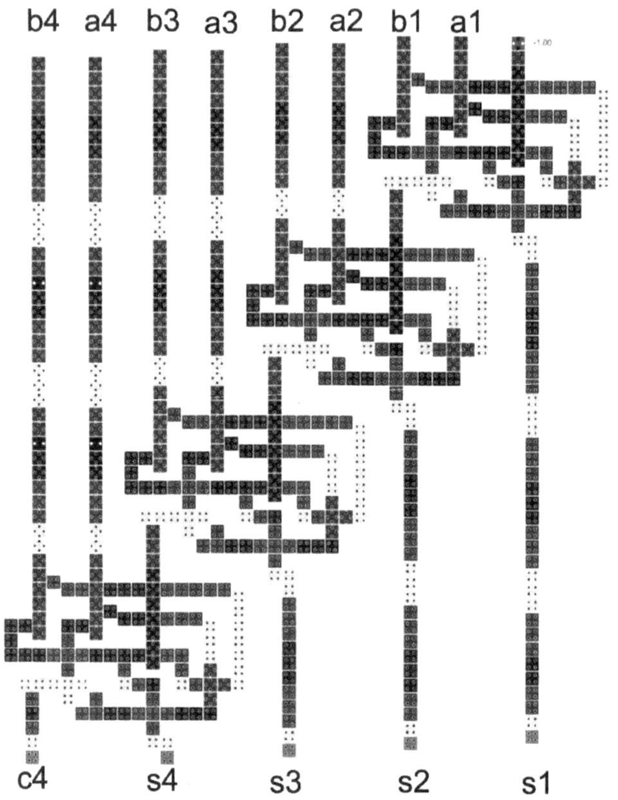

Figure 8.6 Layout of a 4-bit Hänninen adder. (© 2012 IEEE. *From* [37].)

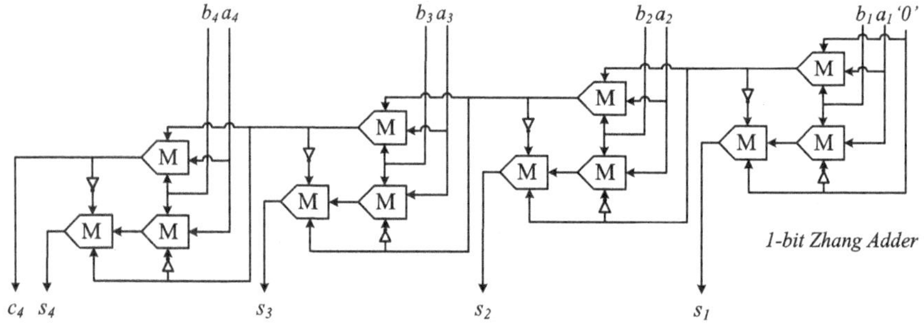

Figure 8.7 Schematic of a 4-bit Zhang adder. (© 2012 IEEE. *From* [37].)

adder, the number of majority gates, inverters, and crossovers are $3n$, $2n$, and $2n$ respectively. The delay of an n-bit Cho adder is $\frac{n+2}{4}$ which can be deduced from its layout or the design results in [14].

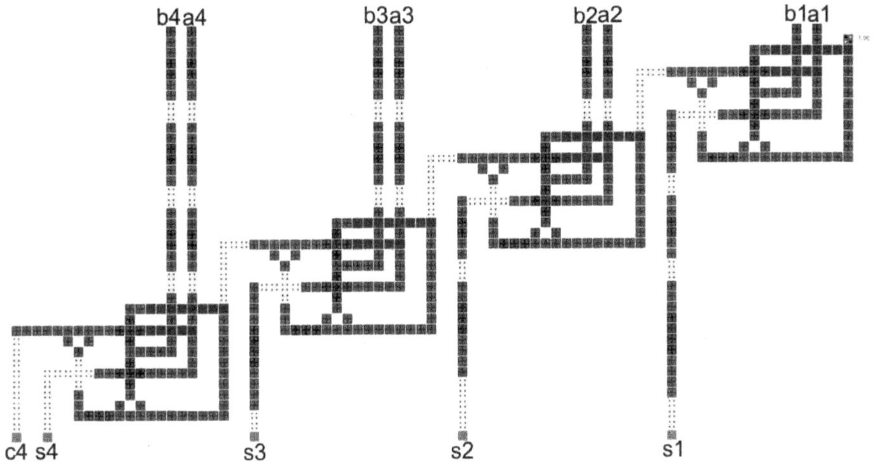

Figure 8.8 Layout of a 4-bit Zhang adder. (© 2012 IEEE. *From* [37].)

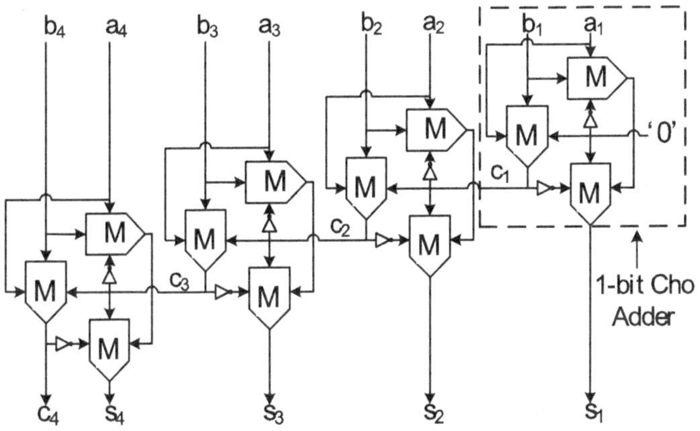

Figure 8.9 Schematic of a 4-bit Cho adder. (© 2012 IEEE. *From* [37].)

8.3.2.3 Pudi Adder

By using some new results in majority logic reduction, several prefix adder designs in QCA such as the Kogge-Stone adder, Brent-Kung adder, Ladner-Fischer adder, and Han-Carlson adder were proposed by Pudi [15]. The QCA Brent-Kung adder performs better than the other QCA prefix adders in terms of delay and area, especially for large adders. Therefore, it is selected as the representative prefix adder in this chapter. The Brent-Kung adder is based on reducing the carry computation to a prefix computation. The addition algorithm of the Pudi adder can be expressed as follows:

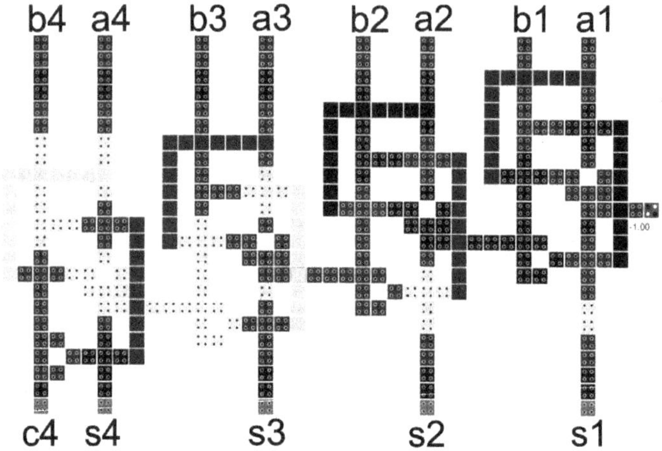

Figure 8.10 Layout of a 4-bit Cho adder. (© 2012 IEEE. *From* [37].)

$$s_i = M\left(\overline{c_i}, M\left(g_i, p_i, \overline{c_i}\right), c_{i-1}\right) \tag{8.9}$$

$$c_i = M\left(g_i, p_i, c_{i-1}\right) \tag{8.10}$$

where, $g_i = a_i \cdot b_i$ and $p_i = a_i + b_i$. The schematic and layout of a 4-bit Pudi adder are shown in Figure 8.11 and Figure 8.12, respectively. The number of majority gates and inverters required for an n-bit Pudi adder are $8n - 3\log_2(n) - 4$ and n, respectively [15].

The number of crossovers in a Pudi adder (i.e., a QCA Brent-Kung adder) is mostly dependent on the prefix computation for the carry-out bits. The prefix graph of a 4-, 8-, and 16-bit Pudi adder is shown in Figure 8.13.

The shaded circles in the prefix graph represent the associative operator [15]. From the prefix graph, it can be seen that when the solid lines cross the dashed lines, crossovers are required. This is also supported by the layout presented by Pudi [15]. Based on this observation, the number of crossovers (denoted as Δ_{pn}) in the prefix graph can be derived as follows:

$$n = 4, \Delta_{p4} > (1 \cdot 1) = \left[2^0 \cdot 2^0\right] \tag{8.11}$$

$$\begin{aligned} n = 8, \Delta_{p8} &> (1 \cdot 3 + 2 \cdot 1) + (1 \cdot 1) \\ &= \left[2^0 \cdot \left(2^2 + 2^1 + 2^0\right) + 2^1 \cdot \left(2^1 + 2^0\right) + 2^2 \cdot 2^0\right] \end{aligned} \tag{8.12}$$

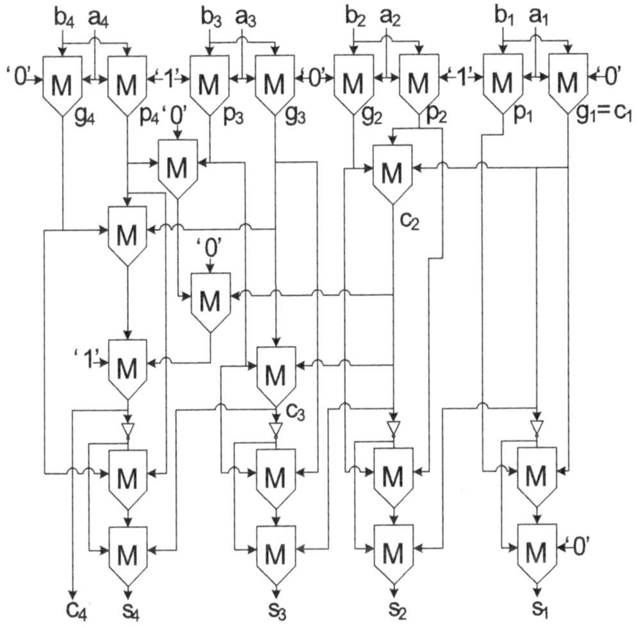

Figure 8.11 Schematic of a 4-bit Pudi adder. (© 2012 IEEE. *From* [37].)

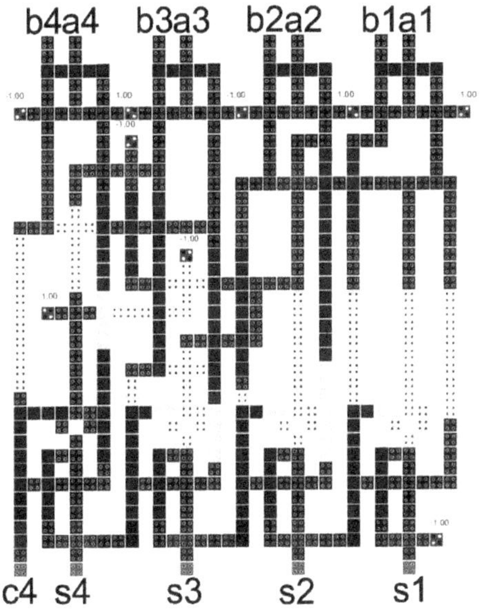

Figure 8.12 Layout of a 4-bit Pudi adder. (© 2012 IEEE. *From* [37].)

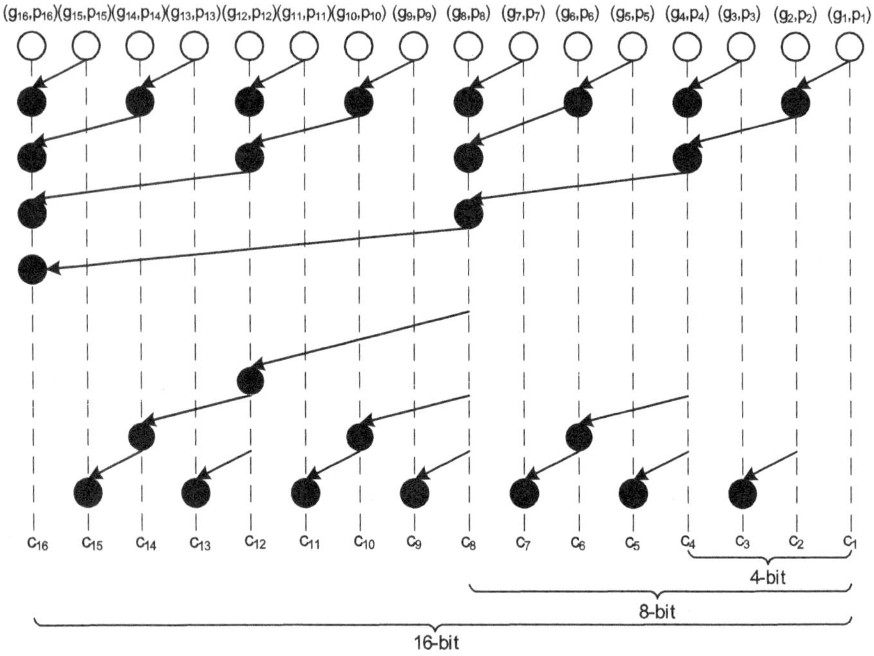

Figure 8.13 16-bit Pudi adder prefix graph.

$$\begin{aligned} n = 16, \Delta_{p16} &> (1\cdot 7 + 2\cdot 3 + 4\cdot 1) + (1\cdot 3 + 3\cdot 1) \\ &= \left[2^0\cdot\left(2^2+2^1+2^0\right)+2^1\cdot\left(2^1+2^0\right)+2^2\cdot 2^0\right] \\ &+ \left[2^0\cdot\left(2^1+2^0\right)+\left(2^1+2^0\right)\cdot 2^0\right] \end{aligned} \tag{8.13}$$

$$\forall n, \Delta_{pn} > \Delta_{pn1} + \Delta_{pn2} \tag{8.14}$$

Let $x = \log_2 n$, where,

$$\begin{aligned} \Delta_{pn1} = &\left[2^0\cdot\left(2^{x-2}+2^{x-3}+\ldots+2^0\right)\right. \\ &\left.+2^1\cdot\left(2^{x-3}+2^{x-4}+\ldots+2^0\right)+\ldots+2^0\right] \end{aligned} \tag{8.15}$$

$$\begin{aligned} \Delta_{pn2} = &\left[\left(2^0-1\right)\cdot\left(2^{x-2}+2^{x-3}+\ldots+2^0\right)+\left(2^1-1\right)\cdot\left(2^{x-3}\right.\right. \\ &\left.\left.+2^{x-4}+\ldots+2^0\right)+\ldots+\left(2^{x-2}-1\right)\cdot 2^0\right] \end{aligned} \tag{8.16}$$

$$\Delta_{pn2} = \Delta_{pn1} - \Big[\left(2^{x-2} + 2^{x-3} + \ldots + 2^0\right) + \left(2^{x-3} + 2^{x-4} + \ldots + 2^0\right) + \ldots + 2^0\Big] \tag{8.17}$$

The above sums, Δ_{pn1} and Δ_{pn2}, can be considered as two progressions. Both are the product progression of a geometric progression and an arithmetic progression. The sum of this kind of progression can be inferred by dislocation subtraction. Therefore,

$$\begin{aligned}\Delta_{pn1} &= 1\cdot 2^0 + 2\cdot 2^1 + 3\cdot 2^2 + \ldots + (x-2)\cdot 2^{x-3} + (x-1)\cdot 2^{x-2} \\ &= (x-2)\cdot 2^{x-1} + 1\end{aligned} \tag{8.18}$$

$$\begin{aligned}\Delta_{pn2} &= \Delta_{pn1} - \left[(x-1)\cdot 2^0 + (x-2)\cdot 2^1 + \ldots + 2\cdot 2^{x-3} + 1\cdot 2^{x-2}\right] \\ &= \Delta_{pn1} - \left(2^x - x - 1\right)\end{aligned} \tag{8.19}$$

$$\begin{aligned}\Delta_{pn} &= 2\cdot\Delta_{pn1} - \left(2^x - x - 1\right) = (x-3)\cdot 2^x + x + 3 \\ &= n\cdot\left(\log_2 n - 3\right) + \log_2 n + 3\end{aligned} \tag{8.20}$$

The above result is for the prefix graph. In addition to these crossovers, there are $3n$ crossovers from the generation of g_i, p_i, and the final sum bits.

Therefore, the total number of crossovers (denoted as C_{pn}) for an n-bit Pudi adder is at least (for a best design):

$$C_{pn} = n\left(\log_2 n - 3\right) + \log_2 n + 3 + 3n \tag{8.21}$$

Next, the delay of an n-bit Pudi adder is derived. There are $2\log_2 n - 1$ stages in the prefix computation. Every stage includes a majority gate delay, which is one clocking zone delay. The final sum stage also has two clocking zone delays. The input and generation of g_i and p_i require two clocking zones. Therefore, the delay (denoted as T_{pn}) of an n-bit Pudi adder is:

$$T_{pn} = \frac{2\log_2 n - 1}{4} + 1 + \text{wire delay} \tag{8.22}$$

Wire delay is introduced in large adders. The long wire (solid line) connecting two remote associative operators should be divided into different clocking zones. The actual wire delay is dependent on the maximum number of cells in a clocking zone. A maximum of 16 cells per clocking zone is assumed in the Pudi adder design [15]. As shown in Figure 8.13, the long horizontal wires crossing the dashed lines will introduce extra delays if they are longer than 16 cells. The number of cells between two dashed lines is four as shown in Figure 8.12. Thus, any long horizontal wire that crosses three dashed lines will introduce one extra clocking zone delay. Therefore, the wire delay (denoted as W_d) can be derived as follows (in units of one clocking zone delay):

$$n = 4,\ W_d = 0 \tag{8.23}$$

$$n = 8,\ W_d = 1 \times 1 = 1 \times 2^0 \tag{8.24}$$

$$n = 16,\ W_d = 1 \times 2 + 2 \times 1 = 2 \times 2^1 \tag{8.25}$$

$$n = 32,\ W_d = 1 \times 4 + 2 \times 2 + 4 \times 1 = 3 \times 2^3 \tag{8.26}$$

$$\forall n,\ W_d = \left(\log_2 n - 2\right) \times 2^{\left(\log_2 n - 3\right)} = \frac{n\left(\log_2 n - 2\right)}{8} \tag{8.27}$$

Thus the delay of an n-bit Pudi adder is:

$$T_{pn} = \frac{2\log_2 n + 3}{4} + \frac{n\left(\log_2 n - 2\right)}{32} \tag{8.28}$$

8.4 Comparison of QCA Adders with Proposed Cost Functions

QCA adders are evaluated in this section with the proposed cost functions. Table 8.1 summarizes all the above adders in terms of the number of majority gates (MG), the number of inverters (INVs), the number of crossings and the delay. The new metrics outlined in Section 8.2 will significantly affect the comparison results in terms of identifying the most suitable adder given the optimization goal, which will be discussed in this section.

Table 8.1
Summary of *N*-bit

n-bit QCA Adders		Number of MGs	Number of INVs	Number of Crossings	Number of Cycles
CP	TA [10]	$5n$	$3n$	$9n$	$n+\frac{1}{4}$
	WA [11]	$3n$	$2n$	$3n$	$n+\frac{1}{4}$
	HA [13]	$3n$	$2n$	$3n$	$n+1$
ML	ZA [12]	$3n$	$2n$	$3n$	n
	CA [14]	$3n$	$2n$	$2n$	$\frac{n+2}{4}$
	PA [15]	$8n-4-3\log_2 n$	n	$n(\log_2 n-3)+3+\log_2 n+3n$	$\frac{2\log_2 n+3}{4}+\frac{n(\log_2 n-2)}{32}$

CP = coplanar, ML = multilayer, TA = Tougaw adder, WA = Wang adder, HA = Hänninen adder, ZA = Zhang adder, CA = Cho adder, PA = Pudi adder

8.4.1 Comparison with Individual Metrics

The six representative adders are first compared in terms of the individual metrics: irreversible power dissipation (M), the fabrication difficulty (C), the delay (T) and complexity ($M + I + C$). The advantages and disadvantages of each adder can be distinguished with each metric. The comparison results for each metric will be helpful for further comparison in terms of the proposed cost functions.

8.4.1.1 Comparison with Complexity

The number of majority gates, inverters and crossovers are used in this research to measure the complexity of a QCA circuit. Assuming the cost of multilayer crossovers is three times that of coplanar crossings, the comparison results are shown in Figure 8.14. The most complex adder is the Pudi adder due to it requiring the largest number of majority gates and crossovers, while Hänninen's coplanar adder is the least complex adder.

8.4.1.2 Comparison with Irreversible Power Dissipation

The number of majority gates is a metric that is related to the irreversible power dissipation as described in Section 8.2. From Figure 8.15, it can be seen that the Pudi adder requires the largest number of majority gates and that the Tougaw adder uses the second largest number of majority gates. The Zhang, Hänninen,

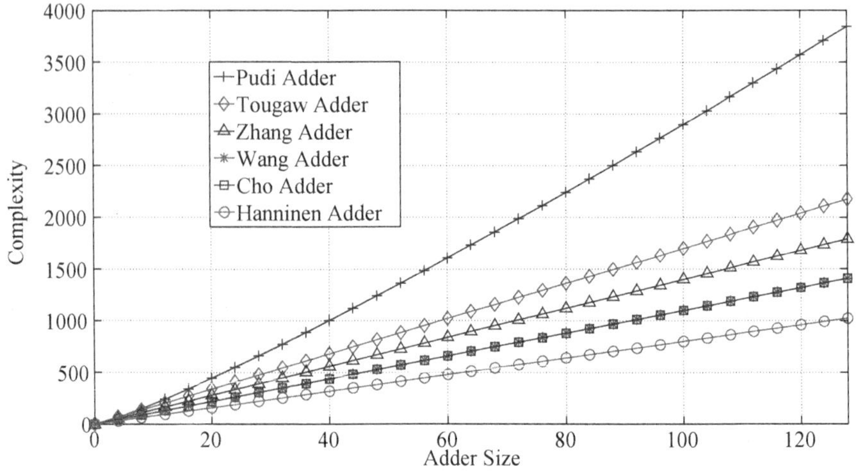

Figure 8.14 Comparison of QCA adders in terms of complexity ($M + I + C$).

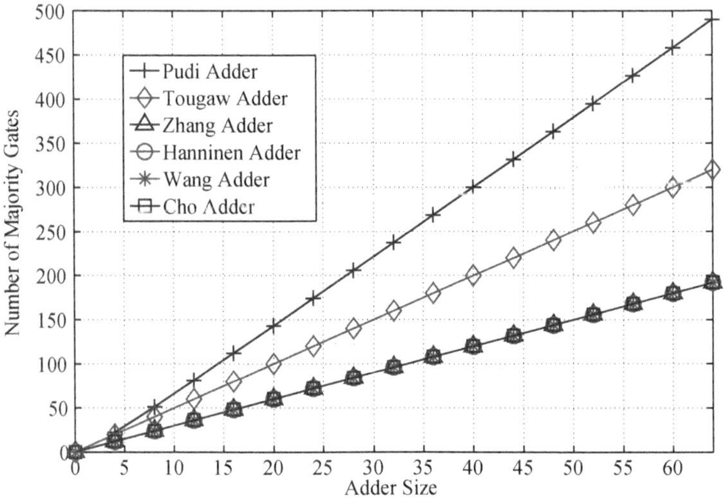

Figure 8.15 Comparison of QCA adders with the meteric of irreversible power dissipation (M).

Wang, and Cho adders use the same and a much smaller number of majority gates.

8.4.1.3 Comparison with Fabrication Difficulty

The number of crossings (C) is used to measure the fabrication difficulty. As mentioned in Section 8.2.4, it is assumed that the cost of multilayer crossovers is three times that of coplanar crossings. From Figure 8.16, it can be seen that

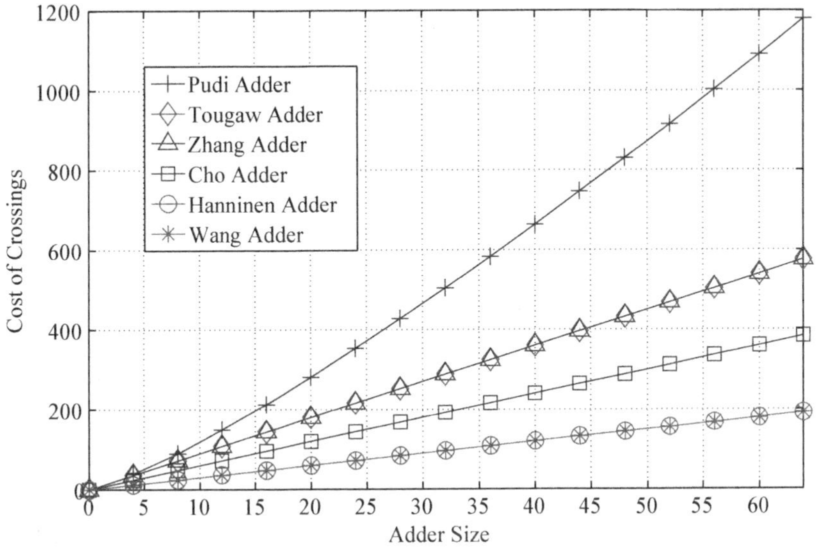

Figure 8.16 Comparison of QCA adders with the metric of cost of crossing assuming that the cost of multilayer crossovers is three times larger than for coplanar crossings.

the multilayer Pudi adder has the highest cost of crossings while Hänninen's coplanar adder has the lowest cost of crossings.

8.4.1.4 Comparison with Delay

Delay is used to measure the speed of a QCA adder. From Figure 8.17, it is evident that the fastest QCA adder is the Pudi adder while the second fastest, Cho's adder, is quite close in terms of speed. The rest of the adders have almost the same delay.

8.4.2 Comparison with QCA Cost Function I

The adders are now compared with a variety of proposed cost functions. As the number of majority gates is associated with both complexity and irreversible power dissipation, a double weighting is applied to M (i.e., $k = 2$). This is also applied to C (i.e., $l = 2$), as the number of crossovers is associated with complexity and fabrication difficulty. Therefore, in most general case, the following cost function can be applied:

$$\text{Cost}_I = \left(M^2 + I + C^2\right) \times T \tag{8.29}$$

The cost difference between multilayer crossovers, C_{ml}, and coplanar crossing, C_{cp}, is considered for two cases:

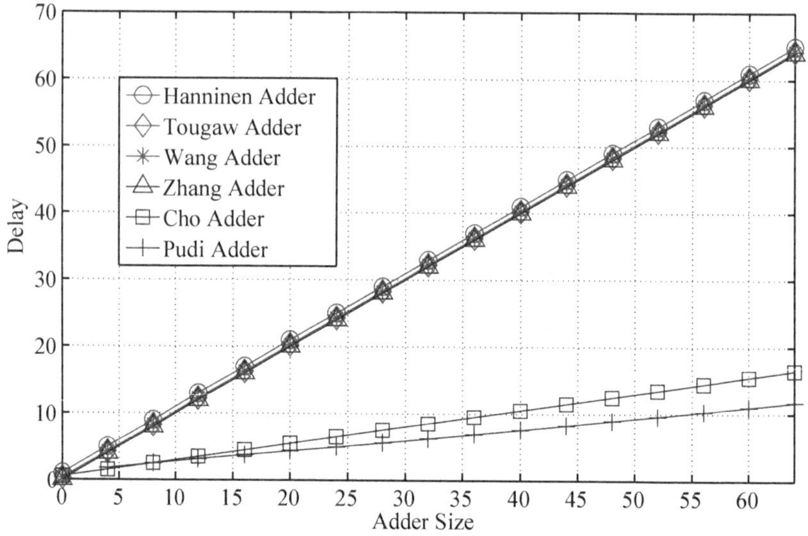

Figure 8.17 Comparison of QCA adder with the metric of delay (T).

8.4.2.1 $C_{ml} = m \times C_{cp}$, $m = 3$

The first case assumes that the cost of multilayer crossovers is 3 times larger than that of coplanar crossings. As shown in Figure 8.18, for this cost function the worst adder is the Tougaw adder as this design is not optimized in terms of complexity and delay. The best adder is the Cho adder due to its small number

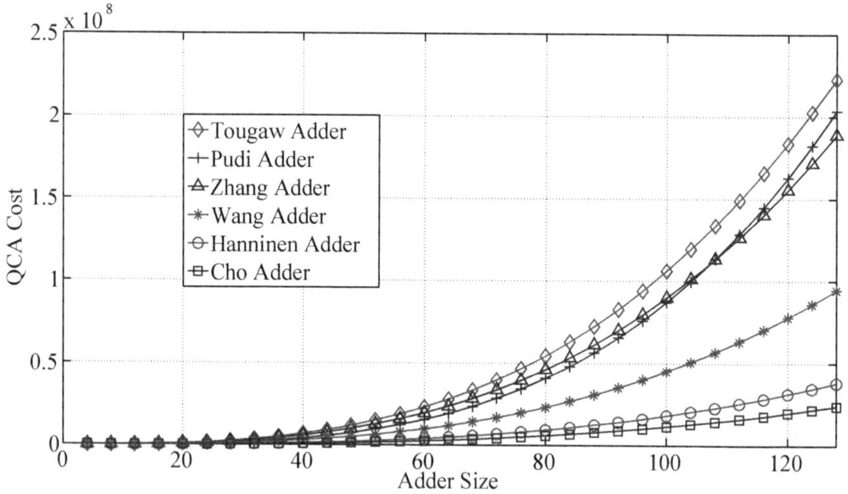

Figure 8.18 Comparison of QCA adder with cost = $(M^2 + I + C^2) \times T$ assuming that the cost of multilayer crossovers is three time larger than for coplanar crossings.

of majority gates, small cost of crossings, and low delay. Although the Pudi adder is found to be the best QCA adder in terms of previous metrics such as area and delay, its number of majority gates and crossovers increases rapidly with the word size, as illustrated. Hence, the large numbers of crossovers and majority gates used in the Pudi adder make it less favorable in terms of overall cost.

8.4.2.2 $C_{ml} = m \times C_{cp}$, $m = 5$

The second case assumes an even larger cost difference between multilayer and coplanar crossovers. In this case, since the cost of multilayer crossovers is large, it is more likely that coplanar adders will be the preferred choice. For the adders that are considered, once *m* is greater than four the best coplanar design will be better than the best of the multilayer adders. Here, a case of $m = 5$ is shown in Figure 8.19. In this case, the worst 64-bit adder is the Zhang adder which is a multilayer adder. For adder sizes larger than 100-bit, the worst adder becomes the Pudi adder which is also a multilayer adder. The best adder for all sizes is the Hänninen adder which is a coplanar adder. Although the same number of majority gates, inverters and crossovers are used in the Zhang adder and the Hänninen adder, the cost of multilayer crossovers significantly affects the design results. Of interest is the Cho adder; although it is a multilayer adder, its cost is

8.4.3 Comparison with QCA Cost Function II

If high speed is the main concern, a higher weighting can be given to the delay metric and the following cost function can be applied:

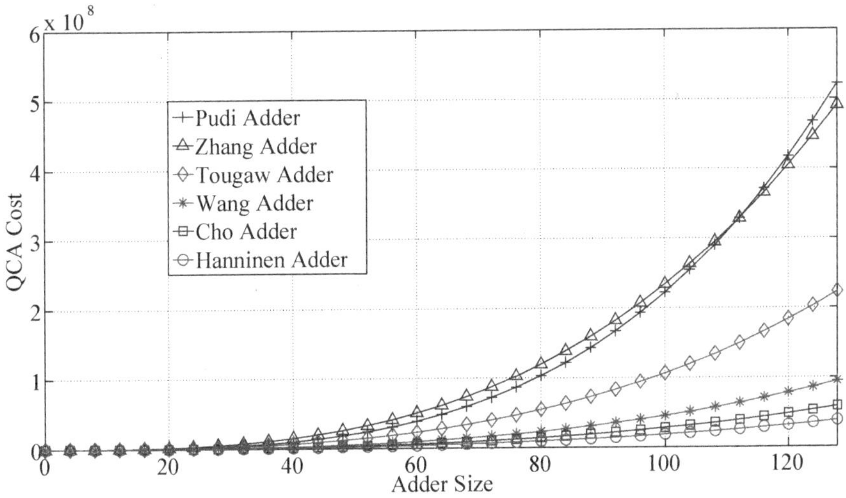

Figure 8.19 Comparison of QCA adders with cost = $(M^2 + I + C^2) \times T$ assuming that the cost of multilayer crossovers is five times larger than for coplanar crossing.

$$\text{Cost}_{II} = \left(M^2 + I + C^2\right) \times T^p \tag{8.30}$$

Note that a double weighting is applied to M and C, (i.e., k=2 and l=2) due to the reason mentioned above. Next, different weightings are considered in relation to performance. Once again, given that the cost of multilayer crossovers is 3 times that of coplanar crossings, the adders are compared in terms of the optimization priority of speed.

8.4.3.1 Comparison with $p = 2$

As shown in Figure 8.20, the Cho adder is the best adder under this cost function. The Pudi adder is the second best and is very close to the performance of the Hänninen adder. Although the Pudi adder is the fastest adder, the cost of its crossovers counteracts its speed advantage to some extent.

8.4.3.2 Comparison with $p = 4$

If a weighting of > 2 is given for speed, the Pudi adder will become the best as it is the fastest. Here, a case of $p = 4$ is shown in Figure 8.21. The best adder now is the Pudi adder, which becomes attractive due to its small delay, while the Cho adder is very close in performance due to its low overall cost. Both are multilayer adders. The worst performing adder is Tougaw's coplanar adder.

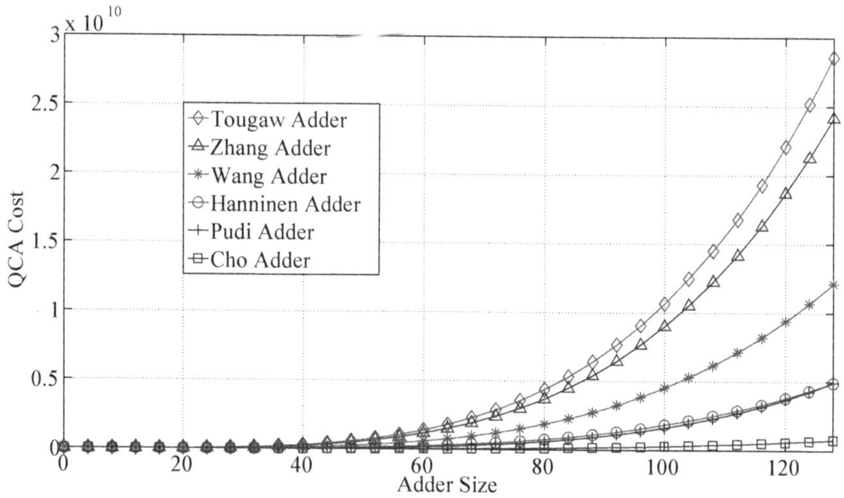

Figure 8.20 Comparison of QCA adders with cost = $(M^2 + I + C^2) \times T^2$ assuming that the cost of multilayer crossovers is three times larger than for coplanar crossings. (© 2012 IEEE. *From* [37].)

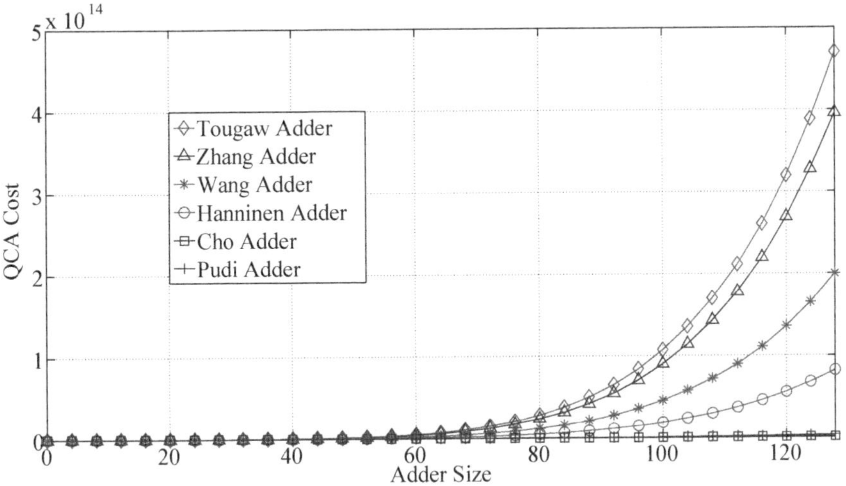

Figure 8.21 Comparison of QCA adders with cost $(M^2 + I + C^2) \times T^4$ assuming that the cost of multilayer crossovers is three times larger than for coplanar crossings.

8.4.4 Discussion

Previous research has compared QCA adders in terms of area, delay, and cell count metrics and shown that multilayer adders were better than coplanar adders due to their higher performance and lower area. However, the area benefits are attained by using high-cost crossings. Important metrics that are unique to QCA technology, such as the cost of crossovers and irreversible power dissipation, are not included in the comparison. By taking the significant difference between the cost of coplanar crossings and that of multilayer crossovers into account, the multilayer adders become less favorable, while the best coplanar adders, such as the Hänninen adder, become more attractive.

The Pudi adder was previously viewed as the best adder in terms of area, cell count, and delay. However, as shown in this work its overall cost is actually very high and may prevent its usage in future QCA systems. Overall, Cho's adder is the best choice for general optimization goals as its overall cost is the lowest in most cases. When speed is the main concern, the best adders are still the multilayer adders. However, the comparison results show that the selection of the most suitable adder is heavily dependent on the optimization goals. In a similar manner to that shown in Section 8.4.3 for speed, irreversible power dissipation, or fabrication cost can be considered as the optimization priorities. For these, different best adders will be found under various optimization goals.

Although these results are based on semiconductor QCA, they can be extended to other implementation technologies. Different cost functions may need to be considered for different implementation technologies. For example, the speed of magnetic QCA [38] is much slower when compared with other QCA implementations. For molecular QCA, speed is not a problem, but the fabrication cost of crossovers is a very significant issue [36]. Therefore, higher weightings should be given to delay (T^p, $p \geq 2$) for magnetic QCA and to the number of crossovers (C^l, $l \geq 2$) for molecular QCA. Therefore, the general cost function proposed, Cost = $(M^k + I + C^l) \times T^p$, can not only be used for different design optimizations but also for different implementation technologies to achieve the best design result. Note that the area metric is not included in the proposed cost functions as it has a lower priority in nanoscale integration. However, the area metric may need to be considered in the cost function for some resource-constrained applications.

8.5 Conclusion

Several cost metrics specific to QCA circuits have been reviewed and discussed in this chapter. Based on the analysis, the delay, the number of majority gates and the number of crossovers are believed to be important elements of a QCA design cost function. A family of cost functions was proposed, Cost = $(M^k + I + C^l) \times T^p$, to evaluate QCA circuits. Adder designs are selected as representative QCA designs. Both coplanar and mulilayer QCA adders have been reviewed and general formulas for the cost metrics for each adder were derived and summarized. A number of QCA adders were also evaluated with the proposed cost functions. The previous "best" adder in terms of speed is shown to be less attractive when the new metrics are taken into account, and coplanar adders become a more preferable choice. The new metrics significantly affect the comparison results, which show that different design goals lead to different favorable adders. The selection of a suitable adder may also be dependent on the implementation technology. As QCA technology continues to develop, the weightings of different metrics may change accordingly and new metrics may need to be considered important. It is hoped that this work will inspire further thought on developing appropriate cost functions for QCA circuits to help improve future QCA circuit design.

References

[1] Walus, K., et al., "QCADesigner: A Rapid Design and Simulation Tool for Quantum-dot Cellular Automata," *IEEE Transactions on Nanotechnology*, Vol. 3, 2004, pp. 26–31.

[2] Walus, K., G. Jullien, and V. Dimitrov, "Computer Arithmetic Structures for Quantum Cellular Automata," in *Conference Record of the 37th Asilomar Conference on Signals, Systems and Computers*, Vol. 2, 2003, pp. 1435–1439.

[3] Hänninen, I., and J. Takala, "Arithmetic Design on Quantum-dot Cellular Automata Nanotechnology," in *Proceedings of the 8th International Workshop on Embedded Computer Systems: Architectures, Modeling, and Simulation,* 2008, pp. 43–52.

[4] Swartzlander, Jr., E., et al., "Computer Arithmetic Implemented with QCA: A Progress Report," in *Conference Record of the 44th Asilomar Conference on Signals, Systems and Computers,* 2010, pp. 1392–1398.

[5] Frost, S., et al., "Memory in Motion: A Study of Storage Structures in QCA," in *Proceedings of1st Workshop on Non-Silicon Computing,* Vol. 2, 2002, pp. 30–37.

[6] Vankamamidi, V., M. Ottavi, and F. Lombardi, "A Line-Based Parallel Memory for QCA Implementation," *IEEE Transactions on Nanotechnology,* Vol. 4, 2005, pp. 690–698.

[7] Niemier, M., M. Kontz, and P. Kogge, "A Design of and Design Tools for A Novel Quantum Dot Based Microprocessor," in *Proceedings of the 37th Annual Design Automation Conference,* 2000, pp. 227–232.

[8] Walus, K., et al., "Simple 4-bit Processor Based on Quantum-dot Cellular Automata (QCA)," in *Proceedings of the 16th IEEE International Conference on Application-Specific Systems, Architecture Processors,* 2005, pp. 288–293.

[9] Liu, W., et al., "Design of Quantum-dot Cellular Automata Circuits Using Cut-Set Retiming," *IEEE Transactions on Nanotechnology,* Vol. 10, 2011, pp. 1150–1160.

[10] Tougaw, P., and C. Lent, "Logical Devices Implemented Using Quantum Cellular Automata," *Journal of Applied Physics,* Vol. 75, 1994, pp. 1818–1825.

[11] Wang, W., K. Walus, and G. Jullien, "Quantum-Dot Cellular Automata Adders," in *Proceedings of the 3rd IEEE Conference on Nanotechnology,* Vol. 1, 2003, pp. 461–464.

[12] Zhang, R., et al., "Performance Comparison of Quantum-dot Cellular Automata Adders," in *Proceedings of the IEEE International Symposium on the Circuits and Systems,* 2005, pp. 2522– 2526.

[13] Hänninen, I., and J. Takala, "Robust Adders Based Quantum-dot Cellular Automata," in *Proceedings of the IEEE International Conference on Application-specific Systems, Architectures and Processors,* 2007, pp. 391–396.

[14] Cho, H., and E. Swartzlander, Jr., "Adder and Multiplier Design in Quantum-Dot Cellular Automata," *IEEE Transactions on Computers,* Vol. 58, 2009, pp. 721–727.

[15] Pudi, V., and K. Sridharan, "Low Complexity Design of Ripple Carry and Brent-Kung Adders in QCA," *IEEE Transactions on Nanotechnology,* Vol. 11, 2012, pp. 105–119.

[16] Thompson, C., "A Complexity Theory for VLSI," Ph.D. dissertation, Carnegie Mellon University, 1980.

[17] Brent, R. P., and H. T. Kung, "The Area-Time Complexity of Binary Multiplication," *Journal of the ACM,* Vol. 28, No. 3, 1981, pp. 521–534.

[18] Nagendra, C., R. M. Owens, and M. J. Irwin, "Power-Delay Characteristics of CMOS Adders," *IEEE Transactions on Very Large Scale Integration Systems,* Vol. 2, 1994, pp. 377–381.

[19] Mead, C., and M. Rem, "Cost and Performance of VLSI Computing Structures," *IEEE Transactions on Electron Devices,* Vol. 26, 1979, pp. 533–540.

[20] Thompson, C. D., "Area-Time Complexity for VLSI," in *Proceedings of the 7th Annual ACM Symposium on Theory of Computing,* 1979, pp. 81–88.

[21] Liu, W., et al., "Design Rules for Quantum-dot Cellular Automata," in *Proceedings of the IEEE International Symposium on Circuits and Systems,* 2011, pp. 2361–2364.

[22] Sengupta, D., and R. Saleh, "Generalized Power-Delay Metrics in Deep Submicron CMOS Designs," *IEEE Transactions on Computer-Aided Design of Integrated Circuits and Systems,* Vol. 26, 2007, pp. 183–189.

[23] Chandrakasan, A., S. Sheng, and R. Brodersen, "Low-Power CMOS Digital Design," *IEEE Journal of Solid-State Circuits,* Vol. 27, 1992, pp. 473–484.

[24] Cho, H., and E. Swartzlander, Jr., "Adder Designs and Analyses for Quantum-Dot Cellular Automata," *IEEE Transactions on Nanotechnology,* Vol. 6, 2007, pp. 374–383.

[25] Landauer, R., "Irreversibility and Heat Generation in the Computing Process," *IBM Journal of Research and Development,* Vol. 5, 1961, pp. 183–191.

[26] Bérut, A., et al., "Experimental Verification of Landauer's Principle Linking Information and Thermodynamics," *Nature,* Vol. 483, No. 7388, 2012, pp. 187–189.

[27] Lent, C., M. Liu, and Y. Lu, "Bennett Clocking of Quantum-dot Cellular Automata and the Limits to Binary Logic Scaling," *Nanotechnology,* Vol. 17, 2006, pp. 4240–4251.

[28] Hänninen, I., and J. Takala, "Irreversible Bit Erasures in Binary Adders," in *Proceedings of the 10th IEEE Conference on Nanotechnology,* 2010, pp. 223–226.

[29] Zhang, R., et al., "A Method of Majority Logic Reduction for Quantum Cellular Automata," *IEEE Transactions on Nanotechnology,* Vol. 3, 2004, pp. 443–450.

[30] Zhang, R., P. Gupta, and N. Jha, "Majority and Minority Network Synthesis with Application to QCA-, SET-, and TPL-Based Nanotechnologies," *IEEE Transactions on Computer-Aided Design of Integrated Circuits and Systems,* Vol. 26, 2007, pp. 1233–1245.

[31] Walus, K., and G. Jullien, "Design Tools for An Emerging SoC Technology: Quantum-Dot Cellular Automata," *Proceedings of the IEEE,* Vol. 94, 2006, pp. 1225–1244.

[32] Gin, A., P. Tougaw, and S. Williams, "An Alternative Geometry for Quantum-Dot Cellular Automata," *Journal of Applied Physics,* Vol. 85, 1999, pp. 8281–8286.

[33] Walus, K., G. Jullien, and R. Budiman, "QCA Coplanar Wire-Crossing and Multilayer Networks," in *Proceedings of iCore Banff Summit,* 2004.

[34] Orlov, A., et al., "Experimental Demonstration of A Binary Wire for Quantum-dot Cellular Automata," *Applied Physics Letters,* Vol. 74, No. 19, 1999, pp. 2875–2877.

[35] Kummamuru, R., et al., "Operation of A Quantum-dot Cellular Automata (QCA) Shift Register and Analysis of Errors," *IEEE Transactions on Electron Devices,* Vol. 50, 2003, pp. 1906–1913.

[36] Chaudhary, A., et al., "Eliminating Wire Crossings for Molecular Quantum-dot Cellular Automata Implementation," in *Proceedings of the 2005 IEEE/ACM International Conference on Computer-Aided Design,* 2005, pp. 565–571.

[37] Liu, W., et al., "A Review of QCA Adders and Metrics," in *Conference Record of the 46th Asilomar Conference on Signals, Systems and Computers,* 2012, pp. 747–751.

[38] Bernstein, G., et al., "Magnetic QCA Systems," *Microelectronics Journal,* Vol. 36, No. 7, 2005, pp. 619–624.

9

Conclusion and Future Work

Weiqiang Liu, Máire O'Neill, and Earl E. Swartzlander, Jr.

9.1 Conclusion

This book presents novel and important contributions to the research in QCA circuit design. A set of QCA design rules has been compiled for the design of robust QCA circuits. Computer arithmetic illustrates that complex and practical circuits can be designed in QCA. By comparing different designs, the characteristics of QCA and transistor circuits are compared and the trade-offs of QCA designs are explored. Design techniques and methodologies that take into account specific issues associated with QCA technologies have been proposed to achieve optimal designs. A number of novel QCA digital circuit designs including adders, multipliers, dividers and cryptographic circuits have been presented. The main conclusions and contributions of each chapter are summarized as follows:

Chapter 3: QCA-optimized adder designs are presented that are based on conventional adder design approaches. RCAs, CLAs, and CSAs have been designed. From these adder designs, the characteristics of QCA are revealed and the advantages and disadvantages of the QCA designs are explored. A complex design in QCA has more timing overhead than in transistor circuits. Transistor circuits are mainly gate-dominant and a reduced gate depth by increasing complexity results in a smaller delay. However QCA circuits are wire-dominant, and an increased complexity causes longer wires, which bring more delays.

In QCA circuit designs, the CSA is worse than the CLA in terms of performance. In this case, gate depth is not the cause of the difference between the equivalent transistor and QCA circuits. It is as a result of the number of cells being limited within a clock zone in wires. In transistor circuits, the wire delay of a small section like an adder block is relatively small. This is not the case in QCA. The CSA design comprises large MUXs and those MUXs have the same select signals. Due to the long select signal wire, the various MUX inputs see the select signal at different times. The pipeline design uses signal synchronization based on the worst case delay. Thus in QCA, the CSA for large word sizes has much larger delays than the CLA. Based on the unique characteristics of QCA, a new adder design is proposed. The CFA is a layout level optimization in QCA using the ripple carry algorithm. The majority gate simplifies the carry computation to only one gate, and by using proper placement and routing, the delay in the carry chain is minimized. The resulting CFA shows a simple structure, a small area, and a fast computation time. This design is a candidate for the best adder design in QCA.

Two cost-efficient QCA BCD adder designs are proposed. A CFA-based BCD adder is designed using the most efficient binary adder, the CFA. Its correction logic is optimized by saving a binary adder during the correction step. An even faster CLA-based decimal adder is designed by directly calculating the decimal sum. Compared with previous QCA decimal adder designs, both of the designs achieve a better performance in terms of latency and overall cost. The CFA-based BCD adder is the smallest QCA BCD adder with the least number of cells. The CLA-based BCD adder is the fastest and most cost-efficient and is the first QCA decimal adder design proposed to use a CLA architecture. Both designs promise efficient decimal arithmetic for the emerging QCA computing paradigm.

Chapter 4: Parallel multipliers are described. Parallel multipliers based on QCAs were expected to achieve attractive results. The two main classes of parallel multipliers are array multipliers and fast multipliers (i.e., Wallace and Dadda multipliers). In CMOS, the areas of the two classes are similar, but the latencies are proportional to the word size for array multipliers and proportional to the logarithm of the word size for fast multipliers. In QCA, to take advantage of the super-pipelined building blocks, simple array structures are implemented. Unlike CMOS technology, structures that can reduce the wiring overhead are available in QCA technology so that RCAs can be faster than CLAs in QCA. Similarly, array multipliers are faster and have less area than the fast multipliers due to their compatibility with QCA technology.

Chapter 5: Two divider designs are presented in this chapter. One is a digit-recurrent restoring binary divider. To use pipelining, an array structure is implemented with controlled full subtractor cell blocks. It can be easily enlarged by irregular block cells without long data connections, but it is large and slow due to the restoring architecture (recursive division) and synchronization problems. However, by using pipelining and a parallel structure it has a good throughput. A Goldschmidt iterative divider using a data tag method shows that sequential circuits in QCA can be built efficiently without state machines. State machines for QCA often have synchronization problems due to the long delays between the state machines and the units to be controlled. The Goldschmidt divider is much smaller and has a greatly reduced latency in comparison to the restoring array divider.

Chapter 6: A cut-set retiming design procedure is proposed to resolve the timing problems associated with QCA designs. This method can efficiently assign clocking zones for relatively large QCA circuits with complicated timing constraints. Case studies involving both a systolic architecture, (i.e., a Montgomery modular multiplier), and a nonsystolic architecture, (i.e., a S27 Benchmark circuit), show that the proposed cut-set retiming design procedure is a very useful and efficient technique to deal with the design challenges presented by the unique four-phase clocking scheme in QCA technology. They also illustrate that complex cryptographic circuits can be designed in QCA using the proposed design procedure.

Chapter 7: Systolic array designs are investigated with two examples, a matrix multiplier and a Galois field multiplier. These case studies show that the internal delay of QCA circuits is accommodated within the systolic array due to its inherent parallelism. Meanwhile, control logic which is unfavorable in QCA technology is also simplified by employing a systolic architecture. The advantages of systolic arrays are shown to be magnified for QCA design when compared with corresponding CMOS implementations. Therefore, systolic arrays are a very attractive design methodology in QCA.

Chapter 8: New cost metrics specific to QCA circuits are reviewed and analyzed. A new family of cost functions based on the new cost metrics is proposed to evaluate QCA circuits. Representative QCA circuits are compared with the proposed cost functions. The selection of the most optimal QCA design is dependent on the optimization goals set in the proposed cost functions. Interestingly, the previous "best" designs in terms of old metrics are found to be less attractive when the new QCA specific metrics are taken into account. Coplanar designs are shown to be preferable in terms of the wire crossing used in QCA design.

9.2 Future Work

This book mainly focuses on the algorithmic design. For other circuits, the physical aspects of designs for QCA cells and circuits along with the algorithmic aspects should be verified. With those levels of designs, all the nonideal device characteristics and process variations can be researched. Sections 9.2–9.2.1 discuss some possible research directions.

9.2.1 QCA Design Automation Tools

Since QCA design tools currently do not support logic synthesis, auto place and route layout level design requires tremendous effort and time to design circuits. Design automation tools that take into account the characteristics of QCA technology could greatly reduce the design efforts. The design methods developed in this book would be very helpful in the design automation of future QCA circuit design and could be integrated within current tools, such as QCADesigner.

9.2.2 Finite State Machine Design

As discussed in Chapter 5, due to the pipelined nature of QCA technology, complex control logic, such as a finite state machine (FSM) is difficult to design since it involves a large number of switches in its data paths. Therefore, conventional FSM architectures are not preferred in QCA technology. Novel techniques to design FSMs based on the characteristics of QCA are needed. Alternative methods such as the data tag method [1] are also viable approaches to resolve the problem. Future work could investigate general methods to design FSMs or alternatives in QCA.

9.2.3 Reversible Circuit Design

Although the power dissipation of QCA circuits is extremely low, research has shown that QCA cryptographic circuits under typical quasi-adiabatic switching are at risk of attack from power analysis [2]. Therefore, some countermeasures may need to be considered during the design process. It may be worth investigating if existing countermeasures [3] proposed for CMOS architectures would still be valid when applied in QCA or if new strategies for QCA cryptographic circuits need to be developed. Alternatively, as QCA provides a practical approach to implementing reversible designs [4, 5], reversible cryptographic circuits in QCA could be used to defend against power analysis attack. Future work could study reversible circuit design for cryptographic applications based on Bennett clocking.

9.2.4 Decimal Arithmetic

Since the 1950s, binary arithmetic has become the standard number system for electronic computers. While binary computer arithmetic design has been extensively investigated, limited attention has been given to decimal arithmetic. However, new financial, commercial and internet-based applications demand high accuracy decimal floating point computations which cannot be achieved with binary arithmetic. The importance of decimal arithmetic has been recognized and its specification has been included in the recent revision to the IEEE 754 standard [6]. Hence, future work could be performed on the design of decimal arithmetic in QCA technology.

References

[1] Kong, I., E. Swartzlander, Jr., and S. Kim, "Design of A Goldschmidt Iterative Divider for Quantum-Dot Cellular Automata," in *Proceedings of the IEEE/ACM International Symposium on Nanoscale Architectures*, 2009, pp. 47–50.

[2] Liu, W., et al., "Are QCA Cryptographic Circuits Resistant to Power Analysis Attack?" *IEEE Transactions on Nanotechnology*, Vol. 11, 2012, pp. 1239–1251.

[3] Lu, Y., M. O'Neill, and J. McCanny, "Evaluation of Random Delay Insertion against DPA on FPGAs," *ACM Transactions on Reconfigurable Technology and Systems*, Vol. 4, No. 1, 2010, pp. 11:1–20.

[4] Lent, C., M. Liu, and Y. Lu, "Bennett Clocking of Quantum-dot Cellular Automata and the Limits to Binary Logic Scaling," *Nanotechnology*, Vol. 17, 2006, pp. 4240–4251.

[5] Frost-Murphy, S. E., et al., "On the Design of Reversible QDCA Systems," *Sandia National Laboratories Technical Report: SAND 2006-5990*, 2006.

[6] "IEEE Standard for Floating-Point Arithmetic," *IEEE Std. 754-2008*, 2008. [Online]. Available: http://grouper.ieee.org/groups/754/

About the Authors

Weiqiang Liu received his B.Eng. in electronic engineering (information engineering) from Nanjing University of Aeronautics and Astronautics (NUAA), Nanjing, China, and his Ph.D. in electronic engineering from the Queen's University Belfast (QUB), Belfast, United Kingdom, in 2006 and 2012, respectively.

He is currently a research fellow in the Institute of Electronics, Communications and Information Technology, QUB. From 2006 to 2009, he was a postgraudate student and research assistant with the DSP Solution Lab in NUAA from, where he was working on field-programmable gate array, microcontroller and DSP hardware design for signal processing. His research interests include QCA circuit designs, very large scale integration circuit designs for signal processing, and cryptographic applications. He is a member of the IEEE.

Earl E. Swartzlander, Jr., received a B.S. from Purdue University, West Lafayette, Indiana, in 1967, an M.S. from the University of Colorado, Boulder, in 1969, and a Ph.D. from the University of Southern California, Los Angeles, in 1972, all in electrical engineering.

He is a professor of electrical and computer engineering at the University of Texas at Austin. In his current position, he and his students conduct research in computer engineering with emphasis on application-specific processor design, including high-speed computer arithmetic, embedded processor architecture, VLSI technology, and nanotechnology. From 1975 to 1990, he held a variety of positions at TRW including the director of independent research and development in the TRW Defense Systems Group, the manager of the Digital Processing Laboratory in the Electronics and Technology Division, and the manager of the Advanced Development Office in the System Development Division. He is the author of one book, the editor of seven books, and the author or coauthor of over 400 refereed journal papers, book chapters, and conference papers.

Professor Swartzlander was the editor-in-chief of *IEEE Transactions on Computers* from 1990 to 1994 and was the founding editor-in-chief of the *Journal of VLSI Signal Processing.* In addition, he has served as an associate editor for *IEEE Transactions on Computers, IEEE Transactions on Parallel and Distributed Systems,* and *IEEE Journal of Solid-State Circuits.* He has been a member of the board of governors of the IEEE Computer Society (1987–1991), the IEEE Signal Processing Society (1992–1994), and the IEEE Solid-State Circuits Council/Society (1986–1991). He has been a member of the IEEE History Committee (1996–2004), the IEEE Fellows Committee (2000–2003), the IEEE James H. Mulligan, Jr., Education Medal Committee (2007–2011), and the IEEE Awards Planning and Policy Committee (2011–present). He has chaired a number of conferences. He is a life fellow of the IEEE and has been honored with the IEEE Third Millennium Medal, the Distinguished Engineering Alumnus Award from the University of Colorado, the Outstanding Electrical Engineer and Distinguished Engineering Alumnus Awards from Purdue University, and the IEEE Computer Society Golden Core Award.

Máire O'Neill received an M.Eng. with distinction and a Ph.D. in electrical and electronic engineering from Queen's University Belfast, Belfast, United Kingdom, in 1999 and 2002, respectively.

She is currently a chair of information security at Queen's and holds an EPSRC Leadership fellowship to conduct research into next generation data security architectures. She previously held a U.K. Royal Academy of Engineering research fellowship from 2003 to 2008. She has authored a research book and has more than 95 peer reviewed conference and journal publications. Her research interests include hardware cryptographic architectures, lightweight cryptography, side channel attacks and countermeasures, physical unclonable functions, and QCA automata circuit design.

Professor O'Neill was guest editor of the launch issue of *IET Information Security* in 2005 and is currently an editorial board member for the *International Journal of Reconfigurable Computing.* She is an IEEE Circuits and Systems for Communications Technical committee member and was treasurer of the executive committee of the IEEE UKRI Section, 2008 to 2009. She is a senior member of the IEEE. She is a fellow of the Higher Education Academy and a member of the IET and the International Association for Cryptologic Research. She has received numerous awards for her research and in 2007 was named British Female inventor of the year at the British Female Inventors and Innovators Network awards.

Heumpil Cho received a B.S. in electrical engineering and an M.S. in electrical engineering and computer science from Seoul National University, Seoul, Korea, and a Ph.D. degree in electrical and computer engineering from the University of Texas at Austin, Austin, in 1998, 2000, and 2006, respectively.

In 2006, he was with Luminary Micro, Inc., Austin, Texas, where he was working on I/O circuit characterization and modeling. From January 2007 to June 2010, he was a senior engineer at Qualcomm, Incorporated, San Diego, California, where he worked on various projects including CDMA/WCDMA/ LTE/ WiMAX wireless modem chip designs. Since July 2010 he has been with Samsung Electronics in Suwon, South Korea. His research interests include high-speed computer arithmetic algorithms, systolic signal processor and CORDIC processor architectures, VLSI circuit designs, architectures for application-specific signal processing, and applications of arithmetic algorithms on QCA. He is a member of the IEEE.

Seong-Wan Kim received a B.S. in electronic engineering from Hanyang University, Seoul, Korea, an M.S. in electrical and electronic engineering from Yonsei University, Seoul, Korea, and a Ph.D. in electrical and computer engineering from the University of Texas at Austin, Austin, in 2002, 2004, and 2011, respectively.

He worked at the Korea Electronics Technology Institute of the Ministry of Commerce, Industry and Energy of ROK in 2004. He worked at Korea Telecom in 2007 and at Alcatel-Lucent USA Inc (Bell Labs), New Jersey, in 2010.

Inwook Kong received a B.S. in electrical engineering from Yonsei University, Seoul, Korea, in 1995. He received an M. Eng. in electrical engineering with a thesis on signal processing using fuzzy clustering in 1997. He received his Ph.D. in electrical and computer engineering from the University of Texas at Austin, Austin, in 2009.

He has worked in the system LSI division of Samsung Electronics, Korea, where he developed ARM-based mobile SOCs from 1997 to 2005. His research interests are high-speed computer arithmetic algorithms, VLSI circuit designs, application-specific signal processing, and arithmetic algorithms on QCA.

Liang Lu received a B.Sc. in telecommunication engineering and an M.S. in signal processing both from Zhejiang University, Hangzhou, China, in 2001 and 2004, respectively, and a Ph.D. in electronic engineering from QUB, United Kingdom, in 2008.

He is currently a leading design engineer at Imagination Technologies, Kings Langley, United Kingdom. He was a Research Fellow in the Institute of Electronics, Communications and Information Technology, QUB, from 2008 to 2012. His research interests include QCA system design, very large scale integration architecture design in video and cryptographic applications.

Roger Woods received a B.S. in electrical and electronic engineering and a Ph.D. in VLSI architectures for recursive filtering from QUB, United Kingdom, in 1985 and 1990, respectively.

He is a professor at Queen's University Belfast. He has published more than 140 scientific papers, and holds a number of patents in the real-time implementation of digital filters and pattern recognition. He has formed a spin-off company Analytic Engines from his research and for which he currently acts as chief technology officer. His research interests include programmable SoC (PSoC) architectures for telecommunications, DSP applications and digital musical instruments and in design tool flows and methodologies for PSoC.

Prof. Woods is a senior member of the IEEE and a fellow of IET. He is also a member of the advisory board to the IEEE Signal Processing Society Technical Committee on the Design and Implementation of Signal Processing Systems. He has been the General Chair of International Conference on Field Programmable Logic (2001), Applied Reconfigurable Computing (2009, 2011), and has acted as the Special Track Program Chair for the IEEE International Conference on Embedded Computer Systems: Architectures, Modelling and Simulation (2013). He is on the technical program committees of a number of FPGA conferences.

Index

Recent Artech House Titles in the Nanoscale Science and Engineering Series

Xiang Zhang and Kang L. Wang, Series Editors

Fundamentals and Applications of Nanomaterials, Zhen Guo and Li Tan

Introduction to Nanorobotic Manipulation and Assembly, Ning Xi and Guangyong Li

Nanoelectronics: Principles and Devices, Second Edition, Mircea Dragoman and Daniela Dragoman

Nanoscale Communication Networks, Stephen F. Bush

Design of Semiconductor QCA Systems, Weiqiang Liu, Earl E. Swartzlander, Jr., and Máire O'Neill